Beyond Geometry

Classic Papers from Riemann to Einstein

Edited with an Introduction and Notes By

Peter Pesic

St. John's College, Sante Fe, New Mexico

DOVER PUBLICATIONS, INC.

Mineola, New York

Acknowledgements

Henri Cartan for his permission to publish the essay by Élie Cartan.

The Albert Einstein Archives for their permission on behalf of the copyright holder of Einstein's writings, The Hebrew University of Jerusalem, to publish the essays by Albert Einstein that appear here.

Bibliographical Note

Beyond Geometry is a new work, first published by Dover Publications, Inc., in 2007

Library of Congress Cataloging-in-Publication Data

Pesic, Peter
Beyond geometry : classic papers from Riemann to Einstein / edited with an introduction and notes by Peter Pesic.
p. cm.
ISBN 0-486-45350-2 (pbk.)

2006050787

Manufactured in the United States of America
Dover Publications, Inc., 31 East 2nd Street, Mineola, N.Y. 11501

Contents

Beyond Geometry

Introduction

"Only the genius of Riemann, solitary and uncomprehended, by the middle of the last century already broke through to a new conception of space, in which space was deprived of its rigidity and in which its power to take part in physical events was recognized as possible."[1] Thus Albert Einstein paid homage to the man he considered his predecessor and inspiration in shaping a geometric vision of physics. Few episodes in the history of ideas and science are as thought-provoking as the passage between Georg Friedrich Bernhard Riemann (1826–1866) and Einstein (1875–1955).

This anthology contains some of the essential writings in this story, gathered so as to facilitate study both of their historical and philosophical significance and their import for larger, unresolved questions about the ultimate nature of space-time. Though a number of these papers have appeared elsewhere, they have never appeared all together in this way; a number of them have never before been translated or reprinted or have remained hard to find. I am responsible for all material in square brackets [...] and, unless otherwise noted, for the translations, which I could not have done without the generous help and careful corrections of Ingo Farin (for the German texts) and Basia Miller (for the French texts), whom I sincerely thank. I also thank Michael Friedman, Jeremy Gray, and Gerald Holton for their helpful comments, Kathy Richards for her typographical artistry, and John Grafton for his editorial support over many years.

Several of the papers here were, in fact, the reading material of the "Olympia Academy" formed by Einstein and two of his friends during his patent office days. The Academy's grandiose name was their joking reaction to pompous institutions; they met in cafés to converse and (if they were lucky) dine on boiled eggs. Perhaps this volume might serve as reading for a new Olympia Academy of friends who would like to read some thought-provoking texts and discuss what they meant and what further unexplored implications they still contain.

Our starting point is Riemann's famous 1854 lecture "On the Hypotheses That Lie at the Foundations of Geometry," which arguably "did more to change our ideas about geometry and physical space than any work on the subject since Euclid's *Elements*."[2] This lecture sets the fundamental perspective as well as the tone and level of the papers chosen to follow it here. For Riemann consciously chose to express his ideas with only a single equation, "almost without calculation," as he put it, the better to bring forward their philosophical import.

Accordingly, I have generally chosen papers that, while serious and important, also expressed themselves primarily in ordinary language, rather than in symbolic mathematics. This was not because Riemann was "popularizing" his ideas; he does not intend to dilute them by presenting them without equations. In his time, natural philosophers considered themselves still philosophers; the names "physicist" and "scientist" were only coined in the 1830s and were resisted by figures like Michael Faraday and James Clerk Maxwell. Riemann was following also German traditions that called for a candidate for university teaching to present both a *Habilitationsschrift* (a learned essay even more developed and original than a doctoral dissertation) and a trial lecture (*Habilitationsvorlesung*) addressed to a general university audience, not only specialists, meant to show the candidate's fitness to teach. Thus, the traditions of the German universities (still observed to this day) met Riemann's desire to open his thoughts as widely as possible.

For that occasion, Riemann offered (as was customary) several different topics to show the breadth of his interests. Famously, his teacher Carl Friedrich Gauss passed over the first two (more purely mathematical and related to Riemann's habilitation essay) and instead chose this one. It should not be presumed that Riemann was caught off-guard or unprepared by Gauss's choice; Riemann himself might well have hoped that this particularly deep and difficult topic would be the subject of his discourse before Gauss, the celebrated "prince of mathematicians." From that point of view, the occasion was memorable; Wilhelm Weber (a friend of Gauss and a noted physicist who was one of Riemann's teachers) noted that Gauss praised Riemann warmly and was noticeably agitated by the daring and import of Riemann's lecture.

To assess what so excited Gauss, we need to seek the themes and questions that led to Riemann's lecture, the grounds on which it itself lay. Here several different strands converge, first among them the nature and significance of non-Euclidean geometries. Since antiquity, geometers wondered whether Euclid's parallel postulate (essentially specifying that only one line through a given point could be parallel to a given line) could be replaced by a simpler axiom. They tried to show that a denial of the parallel postulate would lead to contradiction and found (for instance) that this denial implied that the sizes of triangles would depend on their angles and that there would have to be a fundamental length in geometry. But however bizarre such implications seemed, they were not self-contradictory, as Gauss noted. If we are to believe his later statements, at age fifteen Gauss had already realized that, in fact, a perfectly consistent "anti-Euclidean" geometry (as he called it) could be formulated on the denial of the parallel postulate but he kept this discovery to himself, fearing (as he put it) "the yelling of the Boeotians," of yokels.

By this, Gauss may have meant more the scorn of Kantian philosophers than an outcry of the masses. As late as 1871, the Kantian Rudolf Hermann Lotze declared that all non-Euclidean geometry is nonsense. In contrast, Gauss was critical of Kant's influential argument that the truths of geometry were synthetic a priori, meaning that their validity did not stem from experience, but from a synthesis conditioned by the nature of the mind itself.[3] If so, Euclidean geometry was an inescapable correlate of human thought, "hard-wired" into our very brains, to which Gauss may well have thought that his "anti-Euclidean geometry" was a decisive counterexample.

Though Gauss never published his findings, Nicolai Ivanovich Lobachevsky and János Bolyai independently did so about 1829.[4] Even so, in the opinion of Felix Klein, Gauss "still deserves the greatest credit for non-Euclidean geometry because, by the weight of his authority, he first brought this intellectual creation, heavily contested at first, to general attention and ultimately to victory."[5] To be sure, Bolyai was greatly disappointed to learn that Gauss claimed to have anticipated him in the discovery of non-Euclidean geometry, nor did Gauss give him the praise that was his due. This was not because Bolyai was a weakling; though he unwillingly became a cavalry officer (he could not afford to attend university), he once challenged twelve other officers to duels, on the condition they allow him to refresh himself by playing the violin between bouts. Bolyai defeated all twelve. For his part, Lobachevsky died misanthropic and disappointed that Gauss had not appreciated him sufficiently, though privately Gauss did praise him highly. To others (like Ferdinand Carl Schweikart and Franz Adolf Taurinus), Gauss's encouragement helped them to pursue the new geometry, especially his suggestion that astronomical measurements might test the actual geometry of space, which Lobachevsky had also suggested independently.[6] Such tests would show the experiential character of geometry and space, further refuting Kant.

Though Riemann never mentions non-Euclidean geometry in his lecture, he continually speaks about Gauss's theory of curved surfaces.[7] Gauss's extensive work as a surveyor, both practical and theoretical, prepared his seminal *General Investigations of Curved Surfaces* (1827), especially his "remarkable theorem" (*theorema egregium*): a curved surface has an *intrinsic curvature* that remains unchanged even when it is deformed without stretching and is independent of the way the surface may be embedded in three-dimensional space.[8] Drawing this definition from earlier work by Leonhard Euler and proving this crucial theorem, Gauss brought modern differential geometry into being.

Riemann's lecture takes up Gauss's pioneering work by generalizing it from two-dimensional surfaces to "spaces" of n dimensions. Curiously, Kant himself had examined this possibility in his first published work, "Thoughts on the True Estimation of Living Forces" (1747). Examining Leibniz's concept of *vis viva*, "living force" or kinetic energy (as we now call it), Kant notes that "the ground of the threefold dimension of space is still unknown" and therefore suggests that other worlds may exist with different numbers of dimensions:

> If it is possible that there are extensions with other dimensions, it is also very probable that God has somewhere brought them into being; for His works have all the magnitude and manifoldness (*Mannigfaltigkeit*) of which they are capable. Spaces of this kind, however, cannot stand in connection with those of a quite different constitution. Accordingly such spaces would not belong to our world, but must form separate worlds. Although in what precedes I have shown that it is possible that a number of worlds (in the metaphysical sense

> of the term) may exist together, in the considerations now before us we have, as it seems to me, the sole condition under which it would also be probable that a plurality of worlds actually exist. For if only the one kind of space, that which allows of three dimension, is possible, the other worlds, which I locate outside that in which we exist, could (so far as regards space) be connected with our own, the spaces being all of one kind. We should therefore have to ask why God has separated our world from the others, seeing that through their connection He would have given to his work a greater perfection. For the greater the amount of connection, the greater are the harmony and unity in the world; whereas, on the other hand, gaps and discontinuities violate the laws of order and perfection. It is not, therefore, probable that a plurality of worlds exist (although it is in itself possible), unless it be that those many kinds of space, of which I have just spoken, are likewise possible.[9]

This daring argument shows how bold and visionary Kant could be, though he assumed that any such higher-dimensional space would be Euclidean, in contrast to Riemann's non-Euclidean spaces. In this early writing, Kant already makes use of the crucial term *manifold* (*Mannigfaltigkeit*), the very word Riemann used to describe his generalized n-dimensional spaces that could also be curved, in analogy to curved surfaces of two dimensions. Kant continued to use this term in his *Critique of Pure Reason* (1787) to describe the manifold of spatial and also temporal impressions; in the spirit of his youthful work, perhaps he would have been pleased, had he lived so long, to see Riemann generalize it so amazingly. Here many other mathematical developments converged, especially Hermann Grassmann's development of higher dimensional algebra and Riemann's own work on the theory of complex variables, visualized on "Riemann surfaces."[10]

The thread Riemann chooses through the labyrinth of his curved, multidimensional manifolds is the "line element," the distance between two infinitesimally close points. Gauss had assumed that the line element of a surface generalizes the Pythagorean theorem by including quadratic "cross terms" involving different coordinates.[11] Then Gauss gave a decisive criterion to determine when a surface is intrinsically flat: if his intrinsic curvature vanished *in any single coordinate system*, then the surface really was flat.[12]

Riemann generalized these idea by writing his line element as Gauss had showed him, as a series of all the possible quadratic products of infinitesimal coordinates.[13] Riemann heroically devised the n-dimensional analogue of the Gaussian curvature, already an intricate calculation in two dimensions. In the spirit of his lecture, he does not burden his audience with the derivation of what we now call the Riemann curvature tensor, but did include it elsewhere.[14] Riemann noted also that even a curved manifold is *infinitesimally* flat, Euclidean in a infinitesimal neighborhood of any point, just as Gauss had noticed the same property of curved surfaces, which have a flat infinitesimal neighborhood at every point, a tangent plane.

Though Riemann looms large as a mathematician (his conjecture about the zeta function remains a paramount unsolved mathematical problem),[15] we cannot fully appreciate his 1854 lecture without knowing how deeply he was involved in physics, as were many others among his mathematical colleagues. Gauss himself, beside his extensive work in geodesy and surveying, was an eminent astronomer, served as director of the Göttingen Observatory, and was also deeply involved in studies of electricity and magnetism, interests Riemann shared. In 1853, Riemann became the assistant of his former teacher Weber, who did important research in electromagnetism, including the first determination (1856) of the speed of light by measuring the charge of a Leyden jar first in electrostatic, then in electromagnetic units. (Ole Romer had measured this speed in 1676 by astronomical measurements of the time delays of the appearances of Jupiter's satellites due to the time needed for their light to reach earth.) During the year before his 1854 lecture, Riemann devoted much effort to directly practical questions. He puzzled over a mysterious remainder of electric charge that persisted in seemingly discharged Leyden jars and studied "Nobili's Rings," strange colored rings formed on the plates of voltaic batteries.[16] In all, four of the nine papers Riemann published pertain to physics.

By 1858, Riemann reached the important insight that electromagnetic potentials propagate through empty space at the speed of light. His Göttingen lectures of 1859–1863 show that he came close to realizing the electromagnetic theory found during that period by Maxwell, following the physical insights of Faraday.[17] Already in 1850 Riemann had sought a mathematical theory that "extends the elementary laws valid for individual points to events in the continuously filled space actually given to us, without distinguishing whether mechanics, electricity, magnetism, or thermodynamics is being treated," which he called his "principal task," rather than pure mathematics.[18] This intensely serious and shy man considered "daily self-examination before the face of God" as "the main thing in religion," along with his strict sense of duty. Riemann's health was a crucial factor; by 1863 he had to move to Italy to try to cure the pleurisy that killed him in 1866, not yet forty years of age. Maxwell attributed everything to Faraday's insights; Riemann, though aware of Faraday, did not take him to heart as Maxwell did. This may be all the more poignant if, as Klein thought, "Riemann in the field of mathematics and Faraday in the field of physics are parallel."[19]

Throughout, Riemann's intense activity in what we call physics was deeply informed by his interest in what he called *Naturphilosophie*, an important term whose high metaphysical aspirations are inadequately rendered as "natural philosophy."[20] I have included two excerpts from Riemann's unpublished writings, his *Nachlass*, that indicate something of the character and scope of his thoughts about natural philosophy, especially the implications he might have drawn from his 1854 lecture. Though everywhere else in this volume I have only included complete essays, here I felt it necessary to excerpt, mainly because these writings are themselves fragmentary and unfinished, though we know from Riemann's letters that he eventually intended to publish them. These writings have not been translated before into English, other than in brief citations, but they give a new access into his aspirations to unify the known physical forces.

Here Riemann's interest in what he calls the "*geistlich*" includes both the psychological and the spiritual. His idea of a universal *Stoff* or ether falls in a recognizable tradition reaching back through Euler's pioneering attempts to use the ether to unify the treatment of all fundamental forces to Newton's own speculations about "a certain very subtle spirit pervading gross bodies and lying hidden in them" that might explain electricity, light, and even sensation, were there "a sufficient number of experiments to determine and demonstrate accurately the laws governing the action of this spirit."[21] But Riemann takes this concept of "spirit" even further in the direction of the psychic, explicitly calling on the psychology of Johann Herbart, who succeeded Kant in the chair of philosophy at Königsburg. Herbart held that psychic states are the true actuality, not any substantial, indwelling soul, and that "space is the *symbol of the possible community of things standing in a causal relationship*," especially including their motions.[22]

In his vision of an etherial substance flowing through atoms and disappearing in them, Riemann is attempting to put forward a vision of what we (following Faraday and Maxwell) have come to call a *field*, which (like Herbart's psyche) is more a state or process than a static substance. In this way, Riemann's "ether field" was a emergent form of the new concept of a field, so different in its inner nature from that of discrete atoms. His intriguing assertion that "this substance may be thought of as a physical space whose points move in geometrical space" seems to identify empty space with material ether, though still immersed in "geometrical space." His project to unify gravitation and electromagnetism sets forth the dream of a unified field theory that consumed Einstein's later years. Even Riemann's final work on how the ear discriminates frequencies follows the unifying thread of understanding a continuous medium (here the fluid in the cochlea) as the crux of human hearing.[23]

Though delivered in 1854, Riemann's lecture only appeared in print in 1868, after his untimely death. In those intervening years, though Riemann's lecture remained mostly unread, another remarkable polymath turned his attention to these problems from another angle. As a physician, Hermann von Helmholtz (1821–1894) made great advances in the physiology of vision and of hearing, both fields in which he wrote monumental treatises that remain classics to this day; Riemann's work on hearing directly responds to Helmholtz's 1863 work *On the Sensations of Tone*. Helmholtz also was a crucial advocate of the concept of the conservation of energy and did significant work on electrodynamics, seeking to exclude all considerations of *Geist*.[24] During the course of his studies of the physiology of vision, he became interested in the way geometry might be related to the details of human vision and became convinced that geometry must, therefore, be grounded on what he called "the facts of perception."[25]

Though Helmholtz was not a professional mathematician, he brought to light a rich and interesting mathematical problem. To include the possible curvature of space and also the effects of nonrectilinear coordinates, Riemann's line element is quadratic, excluding terms of higher degrees because he judged that "investigation of this more general class would actually require no essentially different principles, but it would be rather time-consuming and throw proportionally little new light on the study of space, especially since the results cannot be expressed geometrically." Helmholtz sought arguments that would validate Riemann's assumption about the general form of the line element.

Helmholtz published his findings in 1866, just after reading Riemann, and his title reflects his response: "On the Factual Foundations of Geometry," versus Riemann's "On the Hypotheses That Lie at the Foundations of Geometry." I have included this brief account of his argument rather than the more extensive account he published in 1868 because this earlier essay has never before been translated, is concise and does not go into mathematical details, but also because those details of Helmholtz's argument are flawed, as Sophus Lie pointed out.[26] There is also a special interest, I think, in reading the first account of a new argument, in comparison to later versions that may blur or soften the initial insight against criticisms.

Helmholtz argued that if rigid bodies exist that can be moved or rotated without altering their size or structure, then we are forced to restrict ourselves to the general quadratic form of the line element that Riemann chose. Lie's more careful mathematical analysis sustained this part of Helmholtz's argument, but not his further argument that space must therefore be flat, as Eugenio Beltrami had earlier pointed out to Helmholtz.[27] In an afterword to his 1866 paper, Helmholtz conceded that space might have constant nonzero curvature, as in spherical or Lobachevskian geometry. Thus, the existence of rigid bodies implies constant (but not necessarily zero) curvature of space, leaving open the possibilities of variable curvature that Einstein later explored, for which the concept of rigidity would come into question.

In 1870, Helmholtz reviewed the larger context of his argument in "The Origin and Meaning of Geometrical Axioms," a classic essay that appears here as it originally appeared in English, rather than the somewhat longer 1876 version often reprinted elsewhere. This is Helmholtz at his best and most characteristic, by comparison with his terse and telegraphic 1866 paper. In this 1870 paper, he draws on the full range of his experience with visual physiology to frame insights that are both striking and new. He enlarges on his fundamental point that geometry rests on facts, not hypotheses and certainly not a priori truths. He instances how things appear when seen by reflection in a spherical mirror, which transforms everything into an apparently non-Euclidean form but also reveals that the ensuing distortion follows its object faithfully. Here, Helmholtz comes very close to the argument Henri Poincaré will later give that non-Euclidean geometry can be regarded merely as Euclid seen in such a distorting mirror (as in Lewis Carroll's *Through the Looking-Glass*), so that every Euclidean proposition matches up line for line with a mirrored non-Euclidean version, thus implicitly verifying that non-Euclidean geometry is just as consistent as Euclidean, no more, no less. Doubtless drawing on his optical experiments, Helmholtz notes that spherical lenses enable our eyes to experience a non-Euclidean world, to which, though we may be initially disoriented, we become accustomed with surprising ease. So much, then, for any notion that that Euclidean geometry is simply "wired" into our brains.

Apart from Helmholtz's arguments, for many years Riemann's lecture remained mostly unnoticed, as Hermann Weyl noted, "with the exception of a solitary echo in the writings of William Kingdon Clifford" (1845–1879). Clifford became a professor at the University College, London, at age twenty-four, the same year he took his degree, and had a brief but brilliant career as a mathematician and a lecturer on scientific studies before his death from tuberculosis at age thirty-three. In 1873, he published the first English translation of Riemann's 1854 lecture and went on to do important work on non-Euclidean geometry, Riemann surfaces, and the generalized noncommutative algebras now called Clifford algebras.[28]

Clifford not only drew attention to Riemann's ideas but suggested surprising implications in his 1870 address "On the Space-Theory of Matter," of which only the abstract (published in 1876) has survived and is included here. Following Riemann, Clifford argued that "small portions of space *are* in fact of a nature analogous to little hills on a surface which is on the average flat," so that "this variation of the curvature of space is what really happens in that phenomenon which we call the *motion of matter*, whether ponderable or etherial." Clifford went so far as to say that "in the physical world nothing else takes place but this variation," perhaps moved by Faraday's radical vision that so-called "empty" space is the real seat of electromagnetic phenomena. Clifford understood matter to be curved, empty space, whereas Einstein's general relativity retained mass-energy as the separately existent source that curves space. Because it resolves the apparent duality of matter and space, Clifford's idea remains attractive, though no viable physical theory has come from it so far.[29] Here may still lie fruitful insights for the future.

Clifford's 1870 lecture remains an enticing fragment; to complement it, I have included a lecture he gave in 1873 that is extended and leisurely, even prolix, in the grand style much enjoyed in that less hurried age. Speaking at the Royal Institution, where Faraday so often lectured, Clifford gives a genial overview of the whole history of non-Euclidean geometry that is a good starting-point for readers new to the subject, who thus might want to start their reading of this anthology with this selection. By proclaiming "what Copernicus was to Ptolemy, that was Lobachevsky to Euclid," Clifford shows his sympathies with the "revolutionary" new order, which he finds a relief from what he calls the "dreary infinities" of Euclidean space. Clifford notes that "if we suppose the curvature to vary in an irregular manner, the effect of it might be very considerable in a triangle formed by the nearest fixed stars; but if we suppose it approximately uniform to the limit of telescopic reach, it will be restricted to very much narrower limits." For him, the question is not whether space is curved but only what is the scale of its curvature, whether that curvature is of the dimensions of the solar system or larger or smaller. Clifford also argues that, if the curvature of space is positive and "if you were to start in any direction whatever, and move in that direction in a perfect straight line," after traveling "a most prodigious distance," perhaps much greater than between the fixed stars we see, "you would arrive at—this place."

This beguiling possibility is the premise of the next paper, "Elementary Theorems Relating to the Geometry of a Space of Three Dimensions and of Uniform Positive Curvature in the Fourth Dimension" (1877) by Simon Newcomb (1835–1909), which has been not been reprinted until now. Though Newcomb was the most honored American scientist of his time, he signed this paper in a German journal listing his address "Washington U. S. of North America," as if his readers might otherwise not understand where he came from. At first an itinerant schoolteacher, Newcomb gradually taught himself astronomy, started working at the Naval Observatory in Washington, D. C., and eventually became superintendent of the Nautical Almanac Office. His extensive observational and theoretical work concerned the motion of the Moon and planets.

The interest of Newcomb's paper lies in its attempt to set forth a specific model of a positively curved space and deduce its elementary consequences, specifically that such a space would be finite in volume. Newcomb concludes that, though the consequences may seem strange, "there is nothing within our experience which will justify a denial of the possibility that the space in which we find ourselves may be curved in the manner here supposed." Thus, measurements even up to the furthest visible star cannot exclude this possibility, implicitly inviting us to go further still to test it. Like Clifford, Newcomb suggests that as we once learned that the earth was curved, not flat, we may find the same about the universe.[30]

What is more, Newcomb played an important role in preparing a crucial test of general relativity. Einstein wrote that Newcomb was "the last of the great masters" who calculated planetary motions, especially the minute but significant effects of perturbations between planets, "so gigantic is this problem that there are but few who can confront its solutions with independence and critical judgment."[31] Newcomb's exacting work showed that, once all Newtonian perturbations are accounted for, the perihelion (point of nearest approach to the sun) of Mercury's orbit still precesses by forty seconds of arc per century, which general relativity could account for. "It was thus," Einstein concludes, "that the theory of relativity completed the work of the calculus of perturbations and brought about a full agreement between theory and experience." So it was that Newcomb, an early friend of the fourth dimension, played his part in its eventual confirmation.

The role of Henri Poincaré (1854–1912) in the formation of relativity theory is far better known. A brilliant and prolific mathematician, he was also deeply involved in many practical concerns of time-coordination and global geodesy.[32] His work on non-Euclidean geometry runs like a colored thread throughout his works. As mentioned earlier, Einstein's Olympia Academy specifically read and discussed Riemann's lecture, Helmholtz, and especially Poincaré's *Science and Hypothesis* (1902), which "profoundly impressed us and kept us breathless for weeks on end," as one of them wrote later.[33] Rather than reprinting this widely-available text, I have instead included Poincaré's original 1891 essay on "Non-Euclidean Geometries" that form the basis for the book that "held them breathless," along with an 1892 letter Poincaré wrote and later incorporated in that book; neither has previously been translated or reprinted in this form. There are some interesting small points of difference with Poincaré's final text (which I will note in due course), though nothing substantively different; I have, however, provided a new translation that is clearer than those available earlier.

For breathtaking lucidity is the hallmark of Poincaré's style, such as his masterly summation of the different possibilities of non-Euclidean geometry and his dazzling demonstration that they are mutually consistent, as if each were a translation of the other, or a reflection in a curved mirror, as Helmholtz had already surmised. Poincaré does not belabor his proof but presents it so engagingly and concisely (a "dictionary" showing the process of translation) that we seem to understand it immediately. He presents his own influential response to this mutual consistency just as trenchantly. If we were to observe the bending of light rays, he asks (anticipating uncannily one of the classical tests of general relativity), must we then interpret it as resulting from the curvature of space? No, replies Poincaré, for instead space might remain quite Euclidean but some ambient field might cause the bending of light. Poincaré insists that no experiment can decide between these alternative accounts and so the choice between them can only come from our arbitrary, merely conventional choice. For his part, he would prefer to retain Euclidean geometry, for simplicity's sake, and add additional forces as needed to explain phenomena. Einstein would have to ponder this powerful argument long and hard.

For another perspective, I have also included "The Most Recent Researches in Non-Euclidean Geometry" (1893) by Felix Klein (1849–1925), another important figure in mathematics at the time. Klein began his studies in experimental physics but soon directed his attention more towards mathematics, though he was torn between them as late as his inauguration as professor of mathematics in Erlangen (1872), aged twenty-three. For that occasion (though not as his formal habilitation lecture), Klein wrote a pamphlet, now famous as the "Erlangen Program,"[34] outlining a new approach to geometry, which he had learned in the course of his close work with Lie: every geometry should be considered as a space on which a particular group of transformations acts to preserve certain essential invariants. Here, the word "group" comes from Évariste Galois, who coined it in the course of his work on the symmetries of polynomial equations: each group gives a general abstract form to a certain symmetry, as the invariance of rigid bodies in Helmholtz's argument depend on the constant curvature of space.[35] Though Klein's program took a long time to be noticed, by the 1890s it had become a potent direction in mathematics and also had deep resonance in the later developments of relativity and quantum theory.

This rallying cry came from Klein's own work on what he called "the so-called non-Euclidean geometries" because he, like Poincaré and Beltrami, had found ways to interrelate and translate between Euclidean and non-Euclidean geometries.[36] Klein found a way to model non-Euclidean geometry *within Euclid-*

ean geometry itself, by renaming (in the spirit of Poincaré's "dictionary") lines inside a fixed unit circle as "straight lines," redefining distance, then showing those "straight lines" behave as Lobachevsky had said.

Klein gave his 1893 lecture, included here, at a Congress of Mathematics held in connection with the World's Fair that year in Chicago. He gives an historical overview on the various perspectives that had emerged thus far, beginning with the discoveries of Bolyai and Lobachevsky, and including also the perspective of projective geometry, which gives another way to view non-Euclidean geometries as "perspectival" views of the Euclidean. Klein then reviews the way Lie had corrected Helmholtz's arguments and also mentions his own recent work about the different *global* geometries that could seem *locally* indistinguishable. Klein (and Clifford before him) considered different large-scale geometries of space, which now could be not merely curved but oriented or joined up in different ways: one could bend a plane and glue it into a cylinder, or perhaps a torus, or even twist it into a Möbius strip (and its analogue in higher dimensions, the Klein bottle). All these lead Klein "to the opinion that our geometrical demonstrations have no absolute objective truth, but are true only for the present state of our knowledge." There may thus be further possibilities we have not yet conceived, so that "we are led in geometry to a certain modesty, such as is always in place in the physical sciences," delicately recalling Riemann's concluding association of geometry with physics.

In his 1898 paper "On the Foundations of Geometry," Poincaré returns to the ideas of Helmholtz, especially the physiological grounds of our geometric judgments. Poincaré takes Helmholtz's ideas even further by incorporating the concept of group as the essential correlate of these fundamental physiological aspects of perception. After showing the power of this approach, Poincaré concludes that "space is a group," very much in the spirit of Klein's Erlangen Program, though without any mention of Klein. Yet Lie is often cited; Poincaré's implication is that this approach to geometry through groups is widely-held intellectual property that Klein may have popularized but that had many other exponents.[37]

These lines of thought converge on Einstein, who read many of these papers already in his youthful days in the Olympia Academy. Einstein's own writings are represented here by five essays spread over about a decade, often overlapping and returning to the same themes. At the risk of some repetition, I have decided to include them all mainly because they give access to a number of his writings that either have been hard to find or never before translated and because every word from his pen has a special interest and significance. Also, by so doing I thought to help students of Einstein follow the development of his ideas, at least of his exposition of them in a more philosophical, less technical context.

For completeness, among these I have also included two well-known and often-reprinted papers but in revised translations that seek to remedy some of the problems I found in earlier translations. These are the first and last of his papers included here, "Geometry and Experience" (1921) and "The Problem of Space, Ether, and the Field in Physics" (1934). Einstein had already written for the general reader *Relativity: The Special and the General Theory* (first published 1916; currently available as Einstein 1961), which remains to this day perhaps the best such presentation, not just because it comes straight from the theory's author but because he has considerable gifts of wit and expository skill. In "Geometry and Experience," Einstein tries to go more deeply into the philosophical problems underlying his theory, especially his contention that the axioms of geometry are "free creations of the human mind." He also wishes to credit Riemann and Helmholtz (who is not named explicitly here) and to address Poincaré's argument that the physical choice between non-Euclidean and Euclidean geometries is purely conventional. Though he concedes that, "*sub specie aeterni*, Poincaré is right, in my opinion," yet Einstein maintains that physics should proceed following Riemannian, not Euclidean, geometry, pursuant to experimental verification of physical consequences of the effects of geometry on matter and light. In his 1916 book, Einstein had already argued that the *universality* of the effects of gravitation strongly indicated that they were due to geometry's universal effect, for other fields would in general interact differently with different kinds of matter (as electromagnetism depends on electric charge, for instance). In the remainder of his essay, Einstein sketches some of the cosmological implications of his ideas, including Mach's principle (that the inertia of nearby bodies is caused by distant matter) and the cosmological constant. In a final section (added later), he gives examples to argue that we can indeed picture a three-dimensional universe that is finite, yet unbounded.[38]

Einstein's 1925 essay on "Non-Euclidean Geometry and Physics," in contrast, has not appeared in print since 1926 and has never before appeared in English. In it, Einstein gives an important acknowledgment that "all propositions of geometry gain the character of assertions about real bodies ... was especially clearly advocated by Helmholtz, and we can add that without him the formulation of relativity theory would have been practically impossible." This essay also contains Einstein's explicit mention of his interest in Riemann's deep speculations that "the theory of electrical elementary quanta could show the illegitimacy of this concept [of rigid bodies] for distances of the atomic order of magnitude."

In 1926, Einstein wrote an encyclopedia article on "Space-Time" that adds to these considerations the crucial issue of the finiteness and constancy of the speed of light, which (as a measure of an invariant velocity) requires that time enter along with space in any geometrical approach to physics. The reason for this reaches back into the foundations of Newtonian physics. If we wish to determine uniquely the trajectory of a planet, using considerations of spatial geometry alone, we immediately face the difficulty that such a trajectory is determined not only by the initial position of the planet but also by its initial velocity. From any given initial position, an infinity of possible trajectories unfold depending on the planet's initial velocity. Thus, if we wish to determine the planet's subsequent motion, we must have information both about its spatial position and its initial rate of change of position, which involves knowing how the planet is moving *in time*. This alone would have doomed any attempt to geometrize physics without involving time.[39]

Riemann does not indicate his awareness of this elementary fact but he expressed the daring thought that "gravitation and light traveling through empty space must be the same" in terms of "motions of this substance" he identifies with physical space.[40] This little-known passage is the earliest statement I have found that gravitation travels (presumably as a wave, though Riemann does not make this explicit) at a finite velocity that is the same as the velocity of light, anticipating Einstein's deduction of this surprising conclusion sixty years later. As he approached his own field equations, Einstein had the advantage of knowing the completed structure of Maxwell's equations, from which the speed of light emerges as constant and (above all) invariant; without knowing these crucial facts, Riemann could scarcely have thought to unify space and time into a single manifold.

In his 1905 formulation of special relativity Einstein does not explicitly weave space and time into a single fabric, though he acknowledges the unprecedented intermixture of space and time as a result of the relativistic transformations that preserve the constancy of the speed of light. Already in 1764, Jean d'Alembert had noted that "a clever acquaintance of mine believes that it is possible to think of time as a fourth dimension, so that the product of time and solidity would in some sense be the product of four dimensions; it seems to me that this idea, while debateable, has certain merits—at the very least the merit of novelty."[41] This idea only took shape in Hermann Minkowski's famous 1908 lecture on "Space and Time."[42] At first Einstein himself resisted this beautiful insight as merely a mathematical strategem, but shortly he saw that it would be essential in realizing his geometric vision of gravity, which now concerned space-time, not space or time alone.

At this point, we need to reconsider the concept of rigid bodies, previously so significant in these arguments. The invariance of the speed of light, limiting the propagation of all physical influences, requires that even in special relativity every physical body must be deformable, always taking a finite time to respond to impact and hence ultimately not rigid even if we neglect all other physical influences (such as the effects of temperature or elasticity). Though Einstein approached gravitation through considering rigid bodies, because he had to invoke a variable curvature of space-time, the arguments of Helmholtz and Lie no longer apply. Einstein came to realize that rigid bodies cannot be taken as ultimate standards to determine physical geometry. He suggested alternative ways to measure distances by using tightly stretched strings, but a better criterion might be light signals, which he used extensively to formulate special relativity. Weyl and others proposed alternative constructions of the space-time metric based only on light rays and freely falling particles.[43]

Einstein's 1930 essay "Space, Ether, and Field in Physics" has not been reprinted since then, perhaps because it was regarded merely as a first draft for his well-known 1934 essay "The Problem of Space, Ether, and the Field in physics," with which this anthology concludes. But the significantly altered title of the 1934 version suggests that the comparison of these two essays may give us clues about Einstein's changing state of mind during those years. In 1928–1929,

Einstein was working on a unified field theory for which he had great hopes, using the basic framework of Riemannian geometry to try to include the electromagnetic field in a fundamental way as a geometric entity (like space-time itself), rather than as part of the stress-energy tensor Einstein used as the source of the geometrical disturbances in his 1916 general relativity. In his mind, this was and remained a makeshift, an incomplete realization of his bolder vision of *both* "matter" and space-time as purely geometrical entities. Einstein had found fault with earlier attempts by Weyl and Arthur Stanley Eddington to make such a unified field theory; the 1929 theory tried to take advantage of a mathematical possibility that had not been needed in general relativity as originally formulated.[44] This involves what sometimes is called "distant parallelism" and the introduction of "torsion" into Riemannian geometry.

In his 1930 essay, Einstein is full of hope for his idea: "The real is conceived as a four-dimensional continuum with a unitary structure of a definite kind (metric and direction).... The material particles are positions of high density without singularity. We may summarize in symbolical language. Space, brought to light by the corporeal object, made a physical reality by Newton, has in the last few decades swallowed ether and time and seems about to swallow also the field and the corpuscles, so that it remains as the sole medium of reality." Yet by 1934 he cut this from his revised essay; he may have realized that this vision was not working out, for he could not even recover Maxwell's equations in a reasonable physical limit, the weak field case. Nevertheless, until his death he remained true to the deeply-held feeling with which he concludes this later version: "But the idea that there were two structures of space independent of each other, the metric gravitational and the electromagnetic, was intolerable to the theoretical spirit. We are driven to the belief that both sorts of field must correspond to a unified structure of space." The following year, he turned again to his *bête noire*, quantum theory, and formulated his strongest objections in the well-known Einstein-Podolsky-Rosen paper (1935). Part of his frustration with the unified field theory was doubtless also his failure to have it explain quantum theory on the basis of a deeper, deterministic theory.

In the course of his struggle, Einstein turned to the eminent mathematician Élie Cartan (1869–1951), who had proved in 1922 that general relativity was the natural, inevitable generalization of Newtonian mechanics and had independently begun to study the possibilities of torsion in Riemannian geometry.[45] The two had an extended correspondence; Einstein placed great hopes on learning from Cartan's mathematical insights, as he had earlier assimilated Riemannian geometry through his friend Marcel Grossmann. During their period of closest contact Cartan wrote "Euclidean Geometry and Riemannian Geometry" (1931), which has never before been reprinted or translated and gives an admirably lucid account of Cartan's understanding of the whole sweep of modern geometry, including a brief introduction to his concept of torsion.[46]

Though Einstein may have failed in his quest, his vision lives on in other forms. The quest for unification led, among other things, to the standard theory of particle physics, to the unification of the weak and electromagnetic forces, to quantum chromodynamics, and continues to inspire the searches for grand unified theories and the alluring prospects of string theory. Cartan's torsion remains important in the search for quantum gravity, as does Riemannian geometry itself.

All this takes us back to Riemann, whom we should not merely admire as a forerunner but whose vision still holds future prospects. His 1854 remarks about the possibilities that the spatial fabric might break down into discrete "quanta" at very small distances have become the focus of much speculation in the quest for a complete theory of quantum gravity. Many physicists now envisage that some radical change in the nature of space-time occurs near the Planck length, $\sqrt{\frac{hG}{2\pi c^3}} = 1.6 \times 10^{-35}$ m, at which point they anticipate something like a discrete manifold, no longer continuous, as Riemann had suggested. Non-relativistic quantum theory implies that we cannot build clocks and measuring rods not just at atomic scales (the smallest fully functional clock is about the size of a small microorganism); relativistic quantum theory draws from the invariance of the speed of light yet more stringent limitations on the possible measurement of physical quantities.[47] Unlike the problem of rigid bodies, which could be circumvented by light rays and falling particles, these quantum limitations seem more fundamental, ultimately connected with the radical indistinguishability of quanta and the infinite-dimensional Hilbert space mandated by quantum theory.[48]

At whatever scale the issue is joined, though, physicists largely subscribe to Riemann's insight that "upon the exactness with which we pursue phenomena into the infinitely small, does our knowledge of their causal connections essentially depend." He, like Gauss and Lobachevsky before him, also directed our attention to astronomical consequences of the varying curvature of space. Most of all, we realize more and more clearly the force of his insight that "either the reality underlying space must form a discrete manifold, or the basis for the metric relations must be sought outside it, in binding forces acting upon it." Both general relativity and the still-undiscovered theories that will correct and transcend it must struggle, in their own ways, with the quest for these "binding forces," though quantum theory remains the touchstone whose existence Riemann did not suspect and Einstein too much resisted.

This leaves us at the threshold of myriad possibilities that challenge us to add, in our turn, our own ideas of what we think lies "at the foundations of geometry." If space-time is not truly ultimate, what other mathematical structures might lie beneath? Loop quantum gravity aspires to build up space-time without presupposing it. String theory incorporates general relativity into a larger theoretical structure, but does it implicitly rest on accepting space-time as a given backdrop or can it lead to other possibilities?[49] Yet we still lack a fundamental understanding even of the entire range of possibilities. Let me pose the larger question: what are all the possible mathematical structures that could give rise to the *appearance* of space-time, as we know it? This mathematical challenge then leads us to reflect on how, physically, we should judge between the possible alternatives. Once we have left behind the remnants of the perceptual matrix of space-time, what will guide us? When human intuition is altogether abandoned, how are we to find our way among unfamiliar mathematical possibilities?[38] Or are those very possibilities still implicitly in thrall to our preconceptions of space-time?

Here we are very much in need of philosophic reflection, in the spirit in which Einstein and his friends practised it together: free and searching examination of the fundamental questions, not only to give us perspective on what has gone before but even more to help us grope forward. I hope this anthology will provide texts that will continue to offer seminal questions for fruitful study, thought, and discussion. Long live the Olympia Academy!

Peter Pesic

Notes

1. Einstein 1934; see below, 190. Einstein was even asked to make an edition of Riemann's 1854 lecture, but declined; see *ECP* 9:236.

2. Gray 2005a, 507.

3. For Kant's treatment of the analytic and the synthetic in mathematics see Kant 1998, 631–634 (B742–B749). For Lotze, see Russell 1956, 93–109, and Laugwitz 1999, 222. For the philosophical context of Riemann's lecture, see Nowak 1989.

4. For the development of non-Euclidean geometries, see Bonola 1955, Fauvel and Gray 1987, 508–540, and Yaglom 1988, 46–70.

5. Klein 1979, 53.

6. For Bolyai and Lobachevsky, see Yaglom 1988, 56–59; Lobachevsky had a distinguished academic career, including serving as rector of the University of Kazan, but was bitterly disappointed that his great mathematical achievements were ignored or ignorantly criticized. For Schweikart and Taurinus, see Bonola 1954, 75–83. For Gauss's suggestion about astronomical tests, see Bottazzini 1994; for Lobachevsky's, see Daniels 1975.

7. It is possible that Riemann never read Lobachevsky or Bolyai, whom he never mentions, but we know that Riemann did check out from the library the volume of the journal in which Lobachevksi first announced his findings; see Scholz 1982, 220–221. For an excellent intellectual biography that includes much information about the context of Riemann's work, see Laugwitz 1999.

8. For further discussion, see Gauss 2005, v–vi; for the intervening history of differential geometry, see Reich 1973, Kramer 1982, 449–473, Portnoy 1983, and Zund 1983. Jeremy Gray has emphasized the precedence of differential geometry over considerations of non-Euclidean geometry in the whole subsequent development stemming from Riemann's work; see Gray 1979, 1989. For other connections with Riemann's work on complex analysis and Riemann surfaces, see Boi 1995a.

9. Kant 1928, 12–13.

10. For the "manifold of intuition," see Kant 1988, 250; for a thorough treatment of the history of the concept of manifold, see Scholz 1980, as well as his 1982a, 1982b, and 1992, Bernardo 1992, and Boi 1992. Torretti 1978, 109, notes there is no evidence Riemann read Grassmann, though "Grassmann anticipates Riemann in his attempt at a general treatment of 'extended quantities'," which still is "not a general theory of manifolds." For multidimensional spaces, see Klein 1979, 155–169, and Yaglom 1988, 71–94. For the relation to Riemann surfaces, see Laugwitz 1999, 230–231. Klein 1979, 157-158, tells the curious story of the physicist F. Zoellner, who about 1870 took the idea that knots could be undone through access to a higher dimension as a challenge to be resolved by spiritualist experiments; see also Wise 1981, 283.

11. Gauss wrote the line element in two dimensions as $ds^2 = E\ dx^2 + 2F\ dx\ dy + G\ dy^2$, generalizing the Pythagorean theorem, $ds^2 = dx^2 + dy^2$, by introducing functions of position E, F, G and also "cross-terms" (here $2Fdx\ dy$, the 2 introduced for reasons of convenience and symmetry because we expect two equal cross terms, $F\ dx\ dy$ and $F\ dy\ dx$).

12. Note that this works even when the coordinates chosen do not take the familiar Pythagorean form; though polar coordinates ($ds^2 = dr^2 + r^2 d\theta^2$) look different from Cartesian ($ds^2 = dx^2 + dy^2$), both these line elements describe a flat plane.

13. See below, 37, note 10, for the modern expression of Riemann's assumption in terms of a *metric tensor*.

14. Riemann gives it in the second part of a paper concerning the solution of problems in heat conduction he submitted to a competition in Paris in 1861, his "Commentatio"; see Riemann 1990, 401–423, and the detailed discussion and translation in Farwell and Knee 1990, who emphasize that Riemann made no explicit connection between this work and the 1854 lecture. Instead, they treat the "Commentatio" as a contribution to what later became known as tensor analysis. Spivak 1999, 2:181–199, presents a translation of a crucial passage from the "Commentatio" and gives a helpful discussion in modern mathematical language (Riemann used a different notation than is now current). Regarding later developments in tensor calculus, see Zund 1983 and Struik 1989.

15. Many books treat the Riemann hypothesis (that the nontrivial zeros of the Riemann zeta function lie on a certain line in the complex plane) and its profound mathematical implications; see Edwards 2001 and Derbyshire 2003.

16. For Weber's work in relation to Riemann, Gauss, and others, see Klein 1979, 19–21, 232–245, and Wise 1981, 276–287; for Weber's determination of the speed of light, see Everitt 1975, 84, and Maxwell 1954, 2:416–417; Riemann refers to this with respect to his own law of electromagnetic propagation in Riemann 1990, 293. Laugwitz 1999, 254–277, gives a fine overview of Riemann's involvement in physics, as does Tazzioli 1993 and 2002, 81–89. For Riemann's work on Nobili's Rings, see Archibald 1991. Ionescu-Pallas and Sofonea 1986/1987 also places Riemann's work in the context of other work in electrodynamics. This connection was already emphasized by Klein (1881), who argued that Riemann's work on Abelian integrals stemmed from his studies of electricity; see Klein 1963, 12–22.

17. See the citations in note 16 and also Everitt 1975, 80–130. Riemann wrote down mathematical expressions of what now would be called the local conservation of electric charge, the Lorentz gauge, and time-retarded potentials.

18. The quote about extending the elementary laws comes from Riemann's close friend Richard Dedekind in Riemann 1990, 577. For the succeeding quotes about Riemann, see Laugwitz 1999, 1–2, 254–255.

19. Klein, "Riemann und seine Bedeutung für die Entwicklung der modernen Mathematik," in Klein 1921–1923, 3:484, as translated and cited by Bottazzini and Tazzioli 1995, 5. For Faraday's approach in relation to the developments leading to Einstein, see Balibar 1992.

20. This term has particular resonance in the German tradition beyond the English "natural philosophy" because such thinkers as G. W. F. Hegel and J. W. von Goethe took up *Naturphilosophie* in explicit reaction to Newtonian science. Though I do think that Riemann was cognizant of this German usage (and in some ways rather influenced by some of its metaphysical predilections), he, as a mathematician, does not reject Newton, though (as the fragments cited below show) he did want to go beyond him. For the larger context, see Caneva 1997. There are several excellent studies of Riemann's *Naturphilosophie*, including Pettoello 1988, Tazzioli 1993, Bottazzini 1994, Bottazzini and Tazzioli 1995, and Bottazzini 1996.

21. For Euler and Newton, see Heimann 1981; Newton's speculations come from the last pages of his *Principia*, Newton 1999, 943–944. See also Speiser 1927, Buchwald 1981, Wise 1981, Bottazzini 1987, Laugwitz 1999, 285–286, who notes that Riemann learned about Euler's conception from Euler's letters (see Euler 1837, 2:82–95).

22. The quote is from Herbart 1850–1851, 5:306, cited in Cahan 1993, 126. Herbart 1897 gives an overview of his psychology; Riemann may well have been struck by Herbart's attempt to give mathematical form to his psychological theories, as in 18–25 (note also his subtitle to the whole work, "An Attempt to Found the Science of Psychology on Experience, Metaphysics, and Mathematics"); Herbart treats concepts of space and time in 129–140, using different series of sense-perceptions as the primal elements. For a careful discussion of Riemann's relation to Herbart, see Scholz 1982, who argues that (*pace* Russell 1956, 62–70) "Herbart influenced Riemann much more in his epistemology and the comprehension of science than in his particular philosophy of space and spatial thinking." See also Torretti 1978, 107–109 (who does not see Herbart's influence in the 1854 lecture), Bottazzini and Tazzioli 1995, Laugwitz 1999, 287–292, Wise 1981, 284–287 (also concerning Fechner), Nowak 1989, and Boi 1995a.

23. Riemann's quote comes from 43, below. The term "ether field" is not Riemann's own but the way Wise 1981, 288–292, describes Riemann's position, remarking also that Riemann's "project was probably the first attempt at a mathematically founded unified field theory" (289). On Riemann's relation to field theory, see also Laugwitz 1999, 221, 284. For Riemann's "Mechanik des Ohres," see Riemann 1990, 370–382, 807–810 (P. Lax on Riemann's work on shock waves).

24. For the date of publication of Riemann's lecture (variously given as 1866, 1867, and 1868), see Nowak 1989, 40, note 18, who argues for 1868. Helmholtz's treatises on vision and hearing are his 1962a and 1954. For helpful essays that situate this work in context, see Torretti 1978, 155–171, Wise 1981, 295–303, DiSalle 1993. For Helmholtz's essay "On the Conservation of Force," see Helmholtz 1962, 186–222. For a broad survey of Helmholtz's place in nineteenth century science, see Cahan 1993.

25. See his essay "The Facts of Perception," in Helmholtz 1971, 366–408, Helmholtz 1977, 115–185, and Ewald 1996, 2:689–727. For his philosophical significance in relation to Kant and his contemporaries, see Russell 1956, 22–25, 70–81, Stein 1977, 21–23, 36–39 (giving helpful mathematical clarification about the Helmholtz-Lie results, especially the distinction between its local and global versions and its relation to Hilbert's Fifth Problem), Richards 1988, 96–103, and DiSalle 1993.

26. The title of Helmholtz's 1868 essay, "On the Facts That Lie at the Foundation of Geometry," is an even more direct riposte to Riemann's "On the Hypotheses That Lie at the Foundation of Geometry"; regarding the meaning of "hypothesis," see Riemann's comment below, 34, note 1. Helmholtz's 1868 essay is translated in Helmholtz 1977, 39–71; 52, note 9, below gives some cautionary remarks to the reader about this translation and also cites Lie's objections, discussed also by Russell 1956, 47–50. For Lie and continuous groups, see Yaglom 1988, 95–110.

27. For the arguments of Beltrami and Lie, see Torretti 1978, 171–179, and Gray 1989, 147–154. Looking beyond Riemannian geometry, the distance function could be a general function $ds = F(x_i, dx_i)$, where x_i are the coordinates of a point, dx_i its infinitesimal separation from the original, and F an arbitrary function. Such a space is now called a Finsler space, by contrast to a Riemann space, in which F is restricted to be the square root of a quadratic function of the dx_i. See Adler, Bazin, and Schiffer 1975, 10. Laugwitz 1999, 241, points out that in such Finsler spaces, we can define a metric tensor $g_{ik} = \frac{1}{2}\frac{\partial^2 F}{\partial x_i \partial x_k}$, which depends both on the point in question and the tangent vector there; see Laugwitz 1977, 140–156.

28. Weyl 1952, 97. For Clifford's translation, see Riemann 1873 or Clifford 1968, 55–71, also included in Ewald 1996, 652–661; I have chosen to include here Michael Spivak's translation, with his kind permission, which seems to me clearer and more intelligible. In 1843, while trying to generalize complex numbers, William Rowan Hamilton devised quaternions, which turned out not to commute (that is, $AB \neq BA$), the first such explicitly noncommutative algebraic quantities. Clifford generalized this further by concentrating on the structure of their noncommutativity in a way that proved to be useful for quantum theory. For the general theme of noncommutativity, see Pesic 2003. For the relation between Riemann and Clifford, see Farwell and Knee 1992; for Clifford's intellectual context, see Richards 1988, 108–114; for Clifford algebras, see Yaglom 1988, 89–94.

29. For Faraday's vision, see Williams 1980 and Pesic 1988–1989; John Archibald Wheeler [1968] argued that Einstein himself would have preferred Clifford's idea of curved empty space, which Wheeler himself tried to realize [1962] in his "geometrodynamics," though his program faces formidable mathematical problems that have not been resolved. See Grünbaum 1973 for a philosophical appraisal of this approach.

30. Note that Newcomb does not mention Clifford in his 1877 paper but does do so in Newcomb 1898, which covers some of the same ground and hence leads us to think that he only read Clifford after 1877.

31. Cited from Brasch 1929, the text of a letter from Einstein explaining to Newcomb's daughter the importance of her father's work for the confirmation of general relativity.

32. For an engaging account of Poincaré's practical activities (and speculations about Einstein's involvement with chronometric problems), see Galison 2003. My account of both emphasizes their complementary absorption in theoretical questions, not presuming these can be accounted for simply by reference to their practical concerns.

33. Quoted from Solovine 1956, viii. For the Olympia Academy, see Holton 1996, 204–207.

34. For the German text, see Klein 1921, 1:460–497; for a translation, see Rowe 1985 (or Klein, 1893) and the excellent commentary in Gray 2005b, Laugwitz 1999, 246–252, Yaglom 1988, 111–124, and Hawkins 1984 and 2000, 34–42, who emphasizes that, despite its later fame, the Erlangen Program was not widely noticed at the time; see also Rowe 1992. For Poincaré's work on manifolds, see Scholz 1980, 268–347, and 1992.

35. For the context of the emergence of the idea of a group from the question of the solvability of equations, see Pesic 2003. In modern terms, a group is defined as an operation (call it "$*$") that is "closed" over a specific set of elements (so that if a and b are any elements of the group, so too $a * b$ is an element). The elements of the group must also include an identity element I (so that $a * I = I * a = a$) and an inverse a^{-1} for every element a (so that $a * a^{-1} = I$). The particular structure of the "multiplication table" for any group encodes its particular symmetry. For a beautiful account of the relation of groups and symmetries, see Weyl 1980.

36. For these models, see Bonola 1955, 238–264, Rosenfeld 1988, 236–246, Gray 1989, 147–154.

37. As Gray 2005b, 551, remarks, "Another mathematician who had done more on the connections between groups and geometry than Klein ever managed was Poincaré. He had almost certainly come to the idea that groups, and groups of transformations in particular, were fundamental mathematical objects independently of Klein." See also Gray 1992.

38. For a larger examination of the imagination in mathematics, see Mazur and Pesic 2005. For the visualization of Euclidean and non-Euclidean geometries, see Reichenbach 1958, 37–92.

39. This insight was brought forward by Adler, Bazin, and Schiffer 1975, 5–7, and also emerges in Rindler 1994, which imagines how Riemann might have found his way to general relativity (at least in the vacuum). See also Lützen 1999 and Miller 1999.

40. From Riemann's *Nachlass*, below, 43.

41. Cited in Rosenfeld 1988, 179–180, which also instances [illegible] introduction of generalized coordinates (not limited to three in number) and Möbius's use of four-dimensional space in his *Barycentric Calculus* (1827), with the proviso that it "cannot be imagined."

42. For Minkowski's lecture, see Lorentz et al. 1923, 75–91, Galison [illegible] and Walter 1999.

43. For Einstein's relation to the problem of rigid bodies, see *ECP* 2:288–290 [151–153], 2:422–425 [246–248], 3:478–480, and the excellent discussions in Reichenbach 1958, 19–24, Friedman 2002, 201–203, and Stein 1977. Laugwitz 1999, 240, notes that tightly stretched (one dimensional) threads suffice, as Einstein seems to note below, 162. For Weyl's alternative approach to space-time measurements using light rays and falling particles, see Ehlers, Pirani, and Schild 1972.

44. In general relativity, the Riemann curvature tensor is built up from a symmetric *affine connection*, $\Gamma^{\mu}_{\alpha\beta} = \Gamma^{\mu}_{\beta\alpha}$, which shows how vectors change when transported and which is crucial to formulating the concept of the derivative in invariant tensor calculus. In this 1929 theory, Einstein incorporated an anti-symmetric affine connection ($\Gamma^{\mu}_{\alpha\beta} = -\Gamma^{\mu}_{\beta\alpha}$) to accommodate the anti-symmetric electromagnetic field tensor ($F_{\alpha\beta} = -F_{\beta\alpha}$).

45. The same result was found by Weyl in 1921, basically that general relativity is the only metric theory of gravitation that is linear in the derivatives of the metric tensor, of no higher than second degree, and reduces to Newtonian physics in the limit of low velocities and mass-energy density; see Weyl 1952, 315–317, and Pesic and Boughn, 2003. For an interesting perspective on earlier attempts to geometrize physics, see Lützen 1989.

46. For their correspondence, see Cartan and Einstein 1979 and also Cartan's further account of such unified field theories in Cartan 1931; Pais 1982, 325–354, gives an excellent overview of Einstein's various attempts at a unified field theory.

47. For the non-relativistic limits, see Salecker and Wigner 1958, Zimmerman 1962, and Pesic 1993; for the relativistic case, see Berestetskiĭ, Lifshitz, and Pitaevskiĭ 1971, 1:1–4, who give the 1930 arguments of Lev Landau and Rudolf Peierls showing that the ordinary Heisenberg uncertainty relation $\Delta q \Delta p \cong \hbar$ is superseded by $\Delta q \cong \hbar/p$, where Δq is the uncertainty in position q, p the momentum, and $\hbar = h/2\pi$, assuming speeds very close to that of light. Where the non-relativistic formula sets only a mutual exclusion between uncertainty of position and momentum, the relativistic limits each separately.

48. For the issue of indistinguishability, see Pesic 2002, 96, 138 (on Hilbert space).

49. See, for example, the excellent anthologies by Callender and Huggett 2001 and Ashtekar 2005b, which contain numerous papers that invite reflection on the contemporary status of these questions. Regarding loop quantum gravity see Ashtekar 2005a; on the issue of space-time and string theory, see Witten 1996.

On the Hypotheses That Lie at the Foundations of Geometry (1854)

Bernhard Riemann

Plan of the investigation

As is well known, geometry presupposes the concept of space, as well as assuming the basic principles for constructions in space. It gives only nominal definitions of these things, while their essential specifications appear in the form of axioms. The relationship between these presuppositions is left in the dark; we do not see whether, or to what extent, any connection between them is necessary, or a priori whether any connection between them is even possible.

From Euclid to Lagrange, the most famous of the modern reformers of geometry, this darkness has been dispelled neither by the mathematicians nor by the philosophers who have concerned themselves with it. This is undoubtedly because the general concept of multiply extended quantities, which includes spatial quantities, remains completely unexplored.[1] I have therefore first set myself the task of constructing the concept of a multiply extended quantity from general notions of quantity. It will be shown that a multiply extended quantity is susceptible of various metric relations, so that space constitutes only a special case of a triply extended quantity.[2] From this, however, it is a necessary consequence that the theorems of geometry cannot be deduced from general notions of quantity, but that those properties that distinguish space from other conceivable triply extended quantities can only be deduced from experience. Thus arises the problem of seeking out the simplest data from which the metric relations of space can be determined, a problem that by its very nature is not completely determined, for there may be several systems of simple data that suffice to determine the metric relations of space; for the present purposes, the most important system is that laid down as a foundation of geometry by Euclid. These data are—like all data—not logically necessary, but only of empirical certainty, they are hypotheses; one can therefore investigate their likelihood, which is certainly very great within the bounds of observation, and afterwards decide on the legitimacy of extending them beyond the bounds of observation, both in the direction of the immeasurably large, and in the direction of the immeasurably small.

I. Concept of an n-fold extended quantity

In proceeding to attempt the solution of the first of these problems, the development of the concept of multiply extended quantity, I feel particularly entitled to request an indulgent hearing, as I am little practiced in these tasks of a philosophical nature where the difficulties lie more in the concepts than in the construction, and because I could not make use of any previous studies, except for some very brief hints on the subject which Privy Councilor Gauss has given in his second memoir on biquadratic residues, in the Göttingen Gelehrte Anzeige and in the Göttingen Jubilee-book, and some philosophical researches of Herbart.[3]

1

Notions of quantity are possible only when there already exists a general concept that admits particular instances. These instances form either a continuous or a discrete manifold, depending on whether or not a continuous transition of instances can be found between any two of them; individual instances are called points in the first case and elements of the manifold in the second. Concepts whose particular instances form a discrete manifold are so numerous that some concept can always be found, at least in the more highly developed languages, under which any given collection of things can be comprehended (and consequently, in the study of discrete quantities, mathematicians could unhesitatingly proceed from the principle that given objects are to be regarded as all of one kind). On the other hand, opportunities for creating concepts whose instances form a continuous manifold occur so seldom in everyday life that color and the position of sensible objects are perhaps the only simple concepts whose instances form a multiply extended manifold.[4] More frequent opportunities for creating and developing these concepts first occur in higher mathematics.

Particular portions of a manifold, distinguished by a mark or by a boundary, are called quanta.[5] Their quantitative comparison is effected in the case of discrete quantities by counting, in the case of continuous quantities by measurement. Measuring involves the superposition of the quantities to be compared; it therefore requires a means of transporting one quantity to be used as a standard for the others. Otherwise, one can compare two quantities only when one is a part of the other, and then only as to "more" or "less," not as to "how much." The investigations that can be carried out in this case form a general division of the science of quantity, independent of measurement, where quantities are regarded, not as existing independent of position and not as expressible in terms of a unit, but as regions in a manifold. Such investigations have become a necessity for several parts of mathematics, e.g., for the treatment of many-valued analytic functions, and the dearth of such studies is one of the principal reasons why the celebrated theorem of Abel and the contributions of Lagrange, Pfaff, and Jacobi to the general theory of differential equations have remained unfruitful for so long. From this portion of the science of extended quantity, a portion which proceeds without any further assumptions, it suffices for the present purposes to emphasize two points, which will make clear the essential characteristic of an n-fold extension. The first of these concerns the generation of the concept of a multiply extended manifold, the second involves reducing position fixing in a given manifold to numerical determinations.

2

In a concept whose instances form a continuous manifold, if one passes from one instance to another in a well-determined way, the instances through which one has passed form a simply extended manifold, whose essential characteristic is that from any point in it a continuous movement is possible in only two directions, forwards and backwards. If one now imagines that this manifold passes to another, completely different one, and once again in a well-determined way, that is, so that every point passes to a well-determined point of the other, then the instances form, similarly, a doubly extended manifold. In a similar way, one obtains a triply extended manifold when one imagines that a doubly extended one passes in a well-determined way to a completely different one, and it is easy to see how one can continue this construction. If one considers the process as one in which the objects vary, instead of regarding the concept as fixed, then this construction can be characterized as a synthesis of a variability of $n + 1$ dimensions from a variability of n dimensions and a variability of one dimension.

3

I will now show, conversely, how one can break up a variability, whose boundary is given, into a variability of one dimension and a variability of lower dimension. One considers a piece of a manifold of one dimension—with a fixed origin, so that points of it may be compared with one another—varying so that for every point of the given manifold it has a definite value, continuously changing with this point. In other words, we take within the given manifold a continuous function of position, which, moreover, is not constant on any part of the manifold. Every system of points where the function has a constant value then forms a continuous manifold of fewer dimensions than the given one. These manifolds pass continuously from one to another as the function changes; one can therefore assume that they all emanate from one of them, and generally speaking this will occur in such a way that every point of the first passes to a definite point of any other; the exceptional cases, whose investigation is important, need not be considered here. In this way, the determination of position in the given manifold is reduced to a numerical determination and to the determination of position in a manifold of fewer dimensions. It is now easy to show that this manifold has $n - 1$ dimensions, if the given manifold is an n-fold extension. By an n time repetition of this process, the determination of position in an n-fold extended manifold is reduced to n numerical determinations, and therefore the determination of position in a given manifold is reduced, whenever this is possible, to a finite number of numerical determinations. There are, however, also manifolds in which the fixing of position requires not a finite number, but either an infinite sequence or a continuous manifold of numerical measurements. Such manifolds form, e.g., the possibilities for a function in a given region, the possible shapes of a solid figure, etc.

II. Metric relations of which a manifold of n dimensions is susceptible, on the assumption that lines have a length independent of their configuration, so that every line can be measured by every other

Now that the concept of an n-fold extended manifold has been constructed, and its essential characteristic has been found in the fact that position fixing in the manifold can be reduced to n numerical determinations, there follows, as the second of the problems proposed above, an investigation of the metric relations of which such a manifold is susceptible, and of the conditions which suffice to determine them. These metric relations can be investigated only in abstract terms, and their interdependence exhibited only through formulas. Under certain assumptions, however, one can resolve them into relations which are individually capable of geometric representation, and in this way it becomes possible to express the results of calculation geometrically. Thus, although an abstract investigation with formulas certainly cannot be avoided, the results can be presented in geometric garb. The foundations of both parts of the question are contained in the celebrated treatise of Privy Councilor Gauss on curved surfaces.[6]

1

Measurement requires an independence of quantity from position, which can occur in more than one way. The hypothesis which first presents itself, and which I shall develop here, is just this, that the length of lines is independent of their configuration, so that every line can be measured by every other.[7] If position-fixing is reduced to numerical determinations, so that the position of a point in the given n-fold extended manifold is expressed by n varying quantities x_1, x_2, x_3, and so forth up to x_n, then specifying a line amounts to giving the quantities x as functions of one variable. The problem then is to set up a mathematical expression for the length of a line, for which purpose the quantities x must be thought of as expressible in units. I will treat this problem only under certain restrictions, and I first limit myself to lines in which the ratios of the quantities dx—the increments in the quantities x—vary continuously; one can then regard the lines as broken up into elements within which the ratios of the quantities dx may be considered to be constant, and the problem then reduces to setting up a general expression for the line element ds at every point, an expression which will involve the quantities x and the quantities dx. I assume, secondly, that the length of the line element remains unchanged, up to first order, when all the points of this line element suffer the same infinitesimal displacement, whereby

I simply mean that if all the quantities dx increase in the same ratio, the line element changes by the same ratio. Under these assumptions, the line element can be an arbitrary homogeneous function of the first degree in the quantities dx that remains the same when all the quantities dx change sign and in which the arbitrary constants are functions of the quantities x. To find the simplest cases, I first seek an expression for the $(n-1)$-fold extended manifolds which are everywhere equidistant from the origin of the line element, i.e., I seek a continuous function of position which distinguishes them from one another. This must either decrease or increase in all directions from the origin; I will assume that it increases in all directions and therefore has a minimum at the origin. Then if its first and second differential quotients are finite, the first order differential must vanish and the second order differential cannot be negative; I assume that it is always positive.[8] This differential expression of the second order remains constant if ds remains constant and increases quadratically when the quantities dx, and thus also ds, all increase in the same ratio; it is therefore = constant $\times ds^2$ and consequently ds = the square root of an everywhere positive homogeneous function of the second degree in the quantities dx, in which the coefficients are continuous functions of the quantities x. In space, if one expresses the location of a point by rectilinear coordinates, then $ds = \sqrt{\sum(dx)^2}$; space is therefore included in this simplest case. The next simplest case would perhaps include the manifolds in which the line element can be expressed as the fourth root of a differential expression of the fourth degree. Investigation of this more general class would actually require no essentially different principles, but it would be rather time-consuming and throw proportionally little new light on the study of space, especially since the results cannot be expressed geometrically; I consequently restrict myself to those manifolds where the line element can be expressed by the square root of a differential expression of the second degree.[9] One can transform such an expression into another similar one by substituting for the n independent variables, functions of n new independent variables. However, one cannot transform any expression into any other in this way; for the expression contains $n\frac{n+1}{2}$ coefficients which are arbitrary functions of the independent variables; by the introduction of new variables one can satisfy only n conditions, and can therefore make only n of the coefficients equal to given quantities.[10] There remain $n\frac{n-1}{2}$ others, already completely determined by the nature of the manifold to be represented, and consequently $n\frac{n-1}{2}$ functions of position are required to determine its metric relations. Manifolds, like the plane and space, in which the line element can be brought into the form $\sqrt{\sum dx^2}$ thus constitute only a special case of the manifolds to be investigated here; they clearly deserve a special name, and consequently, these manifolds, in which the square of the lines element can be expressed as the sum of the squares of complete differentials, I propose to call flat. In order to survey the essential differences of the manifolds representable in the assumed form, it is necessary to eliminate the features depending on the mode of presentation, which is accomplished by choosing the variable quantities according to a definite principle.[11]

2

For this purpose, one constructs the system of shortest lines emanating from a given point; the position of an arbitrary point can then be determined by the initial direction of the shortest line in which it lies, and its distance, in this line, from the initial point.[12] It can therefore be expressed by the ratios of the quantities dx^0, i.e., the quantities dx at the origin of this shortest line, and by the length s of this line. In place of the dx^0 one now introduces linear expressions $d\alpha$ formed from them in such a way that the initial value of the square of the ine element will be equal to the sum of the squares of these expressions, so that the independent variables are: the quantity s and the ratio of the quantities $d\alpha$. Finally, in place of the $d\alpha$ choose quantities x_1, x_2,..., x_n proportional to them, but such that the sum of their squares equals s^2. If one introduces these quantities, then for infinitely small values of x the square of the line element $= \sum dx^2$, but the next order term in its expansion equals a homogeneous expression of the second degree in the $n\frac{n-1}{2}$ quantities $(x_1 dx_2 - x_2 dx_1)$, $(x_1 dx_3 - x_3 dx_1)$,..., and is consequently an infinitely small quantity of the fourth order, so that one obtains a finite quantity if one divides it by the square of the infinitely small triangle at whose vertices the variables have the values $(0, 0, 0, ...)$, $(x_1, x_2, x_3, ...)$, $(dx_1, dx_2, dx_3, ...)$. This quantity remains the same as long as the quantities x and dx are contained in the same binary linear forms, or as long as the two shortest lines from the initial point to x and from the initial point to dx remain in the same surface element, and therefore depends only on the position and direction of that element. It obviously = zero if the manifold in question is flat, i.e., if the square of the line element is reducible to $\sum dx^2$, and can therefore be regarded as the measure of deviation from flatness in this surface direction at this point. When multiplied by $-\frac{3}{4}$ it becomes equal to the quantity which Privy Councilor Gauss has called the curvature of a surface.[13] Previously, $n\frac{n-1}{2}$ functions of position were found necessary in order to determine the metric relations of an n-fold extended manifold representable in the assumed form; hence if the curvature is given in $n\frac{n-1}{2}$ surface directions at every point, then the metric relations of the manifold may be determined, provided only that no identical relations can be found between these values, and indeed in general this does not occur. The metric relations of these manifolds, in which the line element can be represented as the square root of a differential expression of the second degree, can thus be expressed in a way completely independent of the choice of the varying quantities. A similar path to the same goal could also be taken in those manifolds in which the line element is expressed in a less simple way, e.g., by the fourth root of a differential expression of the fourth degree. The line element in this more general case would not be reducible to the square root of a quadratic sum of differential expressions, and therefore in the expression for the square of the line element the deviation from flatness would be an infinitely small quantity of the second order, whereas for the other manifolds it was an infinitely small quantity of the fourth order. This peculiarity of the latter manifolds therefore might well be called flatness in the smallest parts. For present purposes, however, the most important peculiarity of these manifolds, on whose account alone they have been examined here, is this, that the metric relations of the doubly extended ones can be represented geometrically by surfaces and those of the multiply extended ones can be reduced to those of the surfaces contained within them, which still requires a brief discussion.

3

In the conception of surfaces, the inner metric relations, which involve only the lengths of paths within them, are always bound up with the way the surfaces are situated with respect to points outside them. We may, however, abstract from external relations by considering deformations which leave the lengths of lines within the surfaces unaltered, i.e., by considering arbitrary bendings—without stretching—of such surfaces, and by regarding all surfaces obtained from one another in this way as equivalent.[14] Thus, for example, arbitrary cylindrical or conical surfaces count as equivalent to a plane, since they can be formed from a plane by mere bending, under which the inner metric relations remain the same; and all theorems about the plane—hence all of planimetry—retain their validity. On the other hand, they count as essentially different from the sphere, which cannot be transformed into the plane without stretching. According to the previous investigations, the inner metric relations at every point of a doubly extended quantity, if its line element can be expressed as the square root of a differential expression of the second degree, which is the case with surfaces, is characterized by the curvature. For surfaces, this quantity can be given a visual interpretation as the product of the two curvatures of the surface at this point, or by the fact that its product with an infinitely small triangle formed from shortest lines is, in proportion to the radius, half the excess of the sum of its angles over two right angles. The first definition would presuppose the theorem that the product of the two radii of curvatures is unaltered by mere bendings of a surface, the second, that at each point the excess over two right angles of the sum of the angles of any infinitely small triangle is proportional to its area. To give a tangible meaning to the curvature of an n-fold extended manifold at a given point, and in a given surface direction through it, we first mention that a shortest line emanating from a point is completely determined if its initial direction is given. Consequently we obtain a certain surface if we prolong all the initial directions from the given point which lie in the given surface element, into shortest lines; and this surface has a definite curvature at the given point, which is equal to the curvature of the n-fold extended manifold at the given point, in the given surface direction.

4

Before applying these results to space, it is still necessary to make some general considerations about flat manifolds, i.e., about manifolds in which the square of the line element can be represented as the sum of squares of complete differentials.

In a flat n-fold extended manifold the curvature in every direction, at every point, is zero; but according to the preceding investigation, in order to determine the metric relations it suffices to know that at each point the curvature is zero in $n\frac{n-1}{2}$ independent surface-directions. The manifolds whose curvature is everywhere $= 0$ can be considered as a special case of those manifolds whose curvature is everywhere constant. The common character of those manifolds whose curvature is constant may be expressed as follows: figures can be moved in them without stretching. For obviously figures could not be freely shifted and rotated in them if the curvature were not the same in all directions, at all points.[15] On the other hand, the metric properties of the manifold are completely determined by the curvature; they are therefore exactly the same in all the directions around any one point as in the directions around any other, and thus the same constructions can be effected starting from either; consequently, in the manifolds with constant curvature figures may be given any arbitrary position. The metric relations of these manifolds depend only on the value of the curvature, and it may be mentioned, as regards the analytic presentation, that if one denotes this value by α, then the expression for the line element can be put in the form[16]

$$\frac{1}{1+\frac{\alpha}{4}\sum x^2}\sqrt{\sum dx^2}$$

5

The consideration of *surfaces* with constant curvature may serve for a geometric illustration. It is easy to see that the surfaces whose curvature is positive can always be rolled onto a sphere whose radius is the reciprocal of the curvature; but in order to survey the multiplicity of these surfaces, let one of them be given the shape of a sphere, and the others the shape of surfaces of rotation which touch it along the equator. The surfaces with greater curvature than the sphere will then touch the sphere from inside and take a form like the portion of the surface of a ring, which is situated away from the axis; they could be rolled upon zones of spheres with smaller radii, but would go round more than once. Surfaces with smaller positive curvature are obtained from spheres of larger radii by cutting out a portion bounded by two great semi-circles, and bringing together the cut-lines. The surface of curvature zero will be a cylinder standing on the equator; the surfaces with negative curvature will touch this cylinder from outside and be formed like the part of the surface of a ring which is situated near the axis. If one regards these surfaces as possible positions for pieces of surface moving in them, as space is for bodies, then pieces of surface can be moved in all these surfaces without stretching. The surfaces with positive curvature can always be so formed that pieces of surface can even be moved arbitrarily without bending, namely as spherical surfaces, but those with negative curvature cannot.[17] Aside from this independence of position for surface pieces, in surfaces with zero curvature there is also an independence of position for directions, which does not hold in the other surfaces.

III. Applications to space

1

Following these investigations into the determination of the metric relations of an n-fold extended quantity, the conditions may be given which are sufficient and necessary for determining the metric relations of space, if we assume beforehand the independence of lines from configuration and the possibility of expressing the line element as the square root of a second order differential expression, and thus flatness in the smallest parts.

First, these conditions may be expressed by saying that the curvature at every point equals zero in three surface directions, and thus the metric relations of space are implied if the sum of the angles of a triangle always equals two right angles.

But secondly, if one assumes with Euclid not only the existence of lines independently of configuration, but also of bodies, then it follows that the curvature is everywhere constant, and the angle sum in all triangles is determined if it is known in one.

In the third place, finally, instead of assuming the length of lines to be independent of place and direction, one might assume that their length and direction is independent of place. According to this conception, changes or differences in position are complex quantities expressible in three independent units.

2

In the course of the previous considerations, the relations of extension or regionality were first distinguished from the metric relations, and it was found that different metric relations were conceivable along with the same relations of extension; then systems of simple metric specifications were sought by means of which the metric relations of space are completely determined, and from which all theorems about it are a necessary consequence. It remains now to discuss the question how, to what degree, and to what extent these assumptions are borne out by experience. In this connection there is an essential difference between mere relations of extension and metric relations, in that among the first, where the possible cases form a discrete manifold, the declarations of experience are to be sure never completely certain, but they are not inexact, while for the second, where the possible cases form a continuous manifold, every determination from experience always remains inexact—be the probability ever so great that it is nearly exact. This circumstance becomes important when these empirical determinations are extended beyond the limits of observation into the immeasurably large and the immeasurably small; for the latter may obviously become ever more inexact beyond the boundary of observation, but not so the former.

When constructions in space are extended into the immeasurably large, unboundedness is to be distinguished from infinitude; one belongs to relations of extension, the other to metric relations. That space is an unbounded triply extended manifold is an assumption which is employed for every apprehension of the external world, by which at every moment the domain of actual perception is supplemented, and by which the possible locations of a sought-for object are constructed; and in these applications it is continually confirmed. The unboundedness of space consequently has a greater empirical certainty than any experience of the external world. But its infinitude does not in any way follow from this; quite to the contrary, space would necessarily be finite if one assumed independence of bodies from position, and thus ascribed to it a constant curvature, as long as this curvature had ever so small a positive value. If one prolonged the initial directions lying in a surface direction into shortest lines, one would obtain an unbounded surface with constant positive curvature, and thus a surface which in a flat triply extended manifold would take the form of a sphere, and consequently be finite.[18]

3

Questions about the immeasurably large are idle questions for the explanation of Nature. But the situation is quite different with questions about the immeasurably small. Upon the exactness with which we pursue phenomena into the infinitely small, does our knowledge of their causal connections essentially depend. The progress of recent centuries in understanding the mechanisms of Nature depends almost entirely on the exactness of construction which has become possible through the invention of the analysis of the infinite and through the simple principles discovered by Archimedes, Galileo, and Newton, which modern physics makes use of. By contrast, in the natural sciences where the simple principles for such constructions are still lacking, to discover causal connections one pursues phenomenon into the spatially small, just so far as the microscope permits. Questions about the metric relations of space in the immeasurably small are thus not idle ones.

If one assumes that bodies exist independently of position, then the curvature is everywhere constant, and it then follows from astronomical measurements that it cannot be different from zero; or at any rate its reciprocal must be an area in comparison with which the range of our telescopes can be neglected. But if such an independence of bodies from position does not exist, then one cannot draw conclusions about metric relations in the infinitely small from those in the large; at every point the curvature can have arbitrary values in three directions, provided only that the total curvature of every measurable portion of space is not perceptibly different from zero. Still more complicated relations can occur if the line element cannot be represented, as was presupposed, by the square root of a differential expression of the second degree. Now it seems that the empirical notions on which the metric determinations of space are based, the concept of a solid body and that of a light ray, lose their validity in the infinitely small; it is therefore quite definitely conceivable that the metric relations of space in the infinitely small do not conform to the hypotheses of geometry; and in fact one ought to assume this as soon as it permits a simpler way of explaining phenomena.[19]

The question of the validity of the hypotheses of geometry in the infinitely small is connected with the question of the basis for the metric relations of space. In connection with this question, which may indeed still be ranked as part of the study of space, the above remark is applicable, that in a discrete manifold the principle of metric relations is already contained in the concept of the manifold, but in a continuous one it must come from something else. Therefore, either the reality underlying space must form a discrete manifold, or the basis for the metric relations must be sought outside it, in binding forces acting upon it.[20]

An answer to these questions can be found only by starting from that conception of phenomena which has hitherto been approved by experience, for which Newton laid the foundation, and gradually modifying it under the compulsion of facts that cannot be explained by it. Investigations like the one just made, which begin from general concepts, can serve only to insure that this work is not hindered by unduly restricted concepts and that progress in comprehending the connection of things is not obstructed by traditional prejudices.

This leads us away into the domain of another science, the realm of physics, into which the nature of the present occasion does not allow us to enter.

Notes

[Translated by Michael Spivak 1999, 2:151–162, with his generous permission and with slight revisions by the editor. The German text is given in Riemann 1990, 304–319 (identical with the earlier edition in Riemann 1953, 272–287). Note the commentary by Hermann Weyl in Riemann 1990, 740–769, Weyl 1952, 84–102, Weyl 1988, and Torretti 1978, 82–107, who points out that *der Raum* has the principal meaning of "*the* space," the locus of physical bodies and movements, which is the meaning Riemann generally has in mind, reserving "manifold" for more general n-dimensional "spaces." Accordingly, Spivak originally rendered this term as "Space," which I have left in lower case to accord with common usage, trusting also that this distinction is sufficiently clear from the immediate context. For the larger context of this lecture, see the Introduction; there is a full bibliography in Riemann 1990, 869–910. For an excellent modern commentary that unfolds the mathematical content of Riemann's very condensed summary, see Spivak 1999, 2:163–208.]

1. [Riemann does not seem to be aware that there is no explicit concept of space in Greek mathematics; see Jammer 1966, 15–24, and Huggett 1999 (a useful anthology of writings about space). By "multiply extended quantity" he means the generalization of a single Euclidean magnitude to a magnitude characterized by many numbers. The particular case that engages him is that of ordinary space, which could be considered to be a three-fold extended magnitude, since any point can be characterized by three numbers (say, the Cartesian coordinates of that point x, y, z). But it is important that his arguments are

expressed in terms of a more general n-fold extended magnitude, and (by implication) hence of a n-dimensional "space" (called "Riemann space" or "Riemannian geometry" by later writers) that such magnitudes might characterize. For Riemann's relation to topology, see Bottazzini 1977 and Weil 1979a, 1979b. In light of Newton's famous claim (1999, 943) that "I do not feign hypotheses," note Riemann's comment that "the word hypothesis now has a slightly different meaning than in Newton. Today by hypothesis we tend to mean everything which is mentally added to phenomena"; Riemann 1990, 557, cited in Bottazzini 1994, 26.]

2. [By "metric relations" Riemann means the generalized sense of distance that can connect the separate points into a clearly geometric structure, since a notion of distance seems in some way important to geometry. Riemann makes a radical claim already in this page, in asserting that geometry rests on empirical facts, which are not necessary, but are only hypotheses that admit of observational test.]

3. [For Gauss's writings referred to, see Gauss 1973, 2:110, 116, 119; Laugwitz 1999, 225–226, notes that these allusions are all connected with the complex plane, quoting Gauss's assertion there that "strictly speaking, the essential content of the whole argument belongs to a higher, space-independent, domain of the general abstract science of magnitude that investigates combinations of magnitudes held together by continuity. At present, this domain is poorly developed, and one cannot move in it without the use of language borrowed from spatial images." Riemann may well have drawn inspiration from this. Though he never names him, clearly he includes among "the philosophers" Kant. Riemann explicitly mentions Herbart, as discussed in the Introduction; his reticence about Kant may reflect his desire not to excite the ire of powerful Kantians like Rudolf Hermann Lotze, a professor at Göttingen from 1844 to 1881.]

4. [Riemann was writing during a new burst of work on color vision. He may have been aware of Thomas Young's three-receptor theory of color vision (1802), perhaps also of the early work of Hermann von Helmholtz (1852) and Hermann Grassmann (1853); for a helpful collection, see MacAdam 1970, 51–60. In the same year as Riemann's lecture (1854), James Clerk Maxwell took up this color theory and produced the first color photograph (1861); see Everitt 1975, 63–72, and MacAdam 1970, 62–83. Helmholtz later argued that any perceived color can be specified by three quantities (hue, saturation, and luminosity); see his 1866 *magnum opus* on physiological optics Helmholtz 1962, 2:120–146, excerpted in MacAdam 1970, 84–100; Erwin Schrödinger pointed out that this manifold has a non-Euclidean geometry; his paper is included in MacAdam 1970, 134–193. There is no evidence that Riemann foresaw all this, but it is significant that he already discerned that human vision itself provides an immediate example of a manifold; might he also have surmised that it would prove to be non-Euclidean?]

5. [The use of the term "quanta" may seem prescient to the modern reader who is aware of quantum mechanics. Without diminishing the originality and importance of Riemann's discussion of discreteness here, it might be added that the term "quantum" was much more familiar to his German readers, perhaps

since it was used to render an important Aristotelian and Thomistic distinction rendered by the Latin *quantum*, "how much" (see Jammer 1966, 338–339). Note also that Riemann considers that magnitudes may or may not be independent of position. By that he means that the notion of distance may not apply uniformly and identically throughout the manifold. In Euclidean space this does not occur; the distance between points given by the Pythagorean theorem is the same regardless of where in the plane the points are shifted. But Riemann is aware that it is possible that such a homogenous sense of distance might not apply to all possible spaces; in some spaces there might be certain special points or regions near which the distance function is changed. For instance, Poincaré will use a non-uniform distance function to prove the consistency of Lobachevskian and Euclidean geometries; see below, 100–101.]

6. [As he indicated in his introduction, Riemann considers that the mere existence of such manifolds does not settle the question of their metric (distance). That is, given a (possibly curved) surface in a space of arbitrary dimension, what sort of distance-functions ("metric relations") can describe that surface? In the case of a Euclidean plane, the distance function usually given is that of the straight line between two points, whose length is given by the Pythagorean theorem as the square root of the sum of the squares of the components. But imagine that one were given a table of distances between points. Can one tell whether the surface is curved or flat merely by looking at surveying data compiled of points on that surface, without leaving the surface to "look down" on it from a higher dimension? Riemann is helped by Gauss 2005, which shows that for a two-dimensional surface one can indeed construct a number purely from surveying data gathered in the plane that will give the curvature of the surface at any given point. To construct this "Gaussian curvature" Gauss did not at all need to leave the surface and look back on it from the third dimension. This surprising and beautiful result is most important for Riemann, who proposes to generalize it to surfaces in higher-dimensional spaces.]

7. [Riemann specifically defines "length" or "distance" as independent of the position of the line in question. This is important because otherwise there is a troublesome ambiguity about the notion of distance. Consider a meter stick being moved around; suppose it were somehow noted that the length of that stick is not constant. There is a problem: either the distance function is presumed to be uniform and the surface is curved, or the surface is really flat and the distance-function is not uniform. By *assuming* that distance is uniform, Riemann ascribes any change to the distance between two points on a meter stick to the curvature of the space, rather than to a variable sense of distance, and makes a crucial step towards a geometric understanding of the situation, since here geometry and curvature are asserted to be "real." It may help here to anticipate the controversy that followed Riemann's assertion. By turning this argument around, Poincaré will take the opposite tack, arguing that this assertion is only conventional, not "real," since any presumed curvature can be considered to be merely the appearance of the expansion or contraction of the meter sticks instead.]

8. [In order to find the general form that the distance function must take, Riemann assumes (1) continuity and (2) that distance must be unchanged (to first order) if all the quantities dx are equally displaced. The requirement (2) is a kind of principle of relativity, in that it asserts that the distance cannot change if all the coordinates are shifted slightly. At least in the nearby region, the distance between two points should not vary observably if both points are shifted the same amount. The qualification that it is unchanged "to first order" means that the distance cannot have any term like dx in it, but might have in it higher-order terms like $(dx)^2$ or $(dx)^3$, etc. Note that if dx is itself very small, then the higher-order terms are extremely small. To require that a quantity is unchanged in this way is the same as the requirement that a function be a maximum or minimum: one is requiring that one is "at the bottom" of a trough or "at the top" of a peak, where the first derivatives vanish (and hence there is no change to order dx, as stated). So Riemann's requirement of the local invariance of the distance function is tantamount to requiring that the distance function give the *shortest possible distance* between the two points (shortly we shall show why it is not the greatest possible distance). This seems reasonable enough; it only makes explicit what lies behind one's intuitive sense that "distance" means "least distance" and shows what requires that connection.

The requirement that the distance must be a *homogeneous* function of the first degree of the quantities dx means that there is no constant term; if there were, the distance between a point and itself would not be zero, but would be that given constant. The requirement of continuity (1) seems to indicate that the distance between two points approaches zero continuously as the two points become closer and closer. Therefore the constant must be zero. Note that there may be arbitrary constants that do depend on x, though not on dx, and which hence vanish as $x \to 0$. Riemann adds condition (3): the distance must be unchanged if the signs of all the dx are reversed since the distance must be the same from point A to B as from B to A.

Above we noted that the distance function must be either a maximum or a minimum; Riemann notes that it conforms to the basic ordinary sense of distance that it increase as points move apart and hence the distance is a minimum. Because of that, the distance function follows the ordinary requirements for a minimum and must therefore be composed of no first derivative terms (which would make it not a minimum) and only positive second derivatives (negatives would make it a maximum). So the distance function or *line element* can be written as ds^2, where ds^2 depends only on second-derivative terms like dx^2. Note that now condition (3) is naturally satisfied, since all terms dx^2 are unchanged if one reverses $dx \to -dx$. Also ds^2 is inherently positive, which is consistent with the sense that distance is never negative and always increases as points are further removed from each other. Riemann explicitly notes how the Pythagorean distance in the Euclidean plane or in Euclidean space obeys this general form; there, $ds = \sqrt{\sum(dx^2)} = \sqrt{dx^2 + dy^2 + dz^2}$. This example induces him to ignore less simple forms of distance involving fourth roots rather than square roots.]

9. [Helmholtz takes up what would justify limiting the line element in this way, later known as the Helmholtz-Lie space problem; see 47–52 below.]

10. [Riemann suggests that we consider introducing n new independent variables to replace the old ones, with the implicit thought that there should be such a set of new variables that will always unscramble the distance into something very similar to its Pythagorean form. Now he notices that there are conditions implicit in the distance function. He considers all the coefficients that might occur in that function. Let's list a few:

$$ds^2 = g_{11}dx_1dx_2 + g_{12}dx_1dx_2 + g_{21}dx_2dx_1 + g_{22}dx_2dx_2 + \ldots,$$

where the n variables are called $x_1, x_2, x_3, \ldots x_n$ (to give a simpler general notation than using $x, y, z, \ldots$ or $r, \theta, \phi, \ldots$) and the coefficients are called $g_{11}, g_{12}, \ldots$. In general, the coefficient of the term $dx_\sigma dx_\tau$ is called $g_{\sigma\tau}$; this notation is the one Einstein will later use, by summing over the indices σ and τ :

$$ds^2 = \sum_{\tau\sigma} g_{\sigma\tau}dx^\sigma dx^\tau.$$

The $g_{\sigma\tau}$ is often called a *tensor*, to indicate that it is the generalization of a vector. In a n-dimensional space, a vector would have n components and a tensor would have $n \times n$ components. Often $g_{\sigma\tau}$ is specifically called the *metric tensor*, to emphasize that it characterizes the metric (distance) function of the space. A scalar has a magnitude only; a vector has a magnitude and a direction; a tensor of this kind has a magnitude and many directions.

Riemann realizes that the component of dx_1dx_2 must be the same as the component of dx_2dx_1, since the order of the differentials is immaterial. Thus $g_{12} = g_{21}$ and in general $g_{\sigma\tau} = g_{\tau\sigma}$. In the language of tensors, $g_{\sigma\tau}$ is said to be *symmetric*. We can also write the components $g_{\sigma\tau}$ as a matrix:

$$g_{\sigma\tau} = \begin{pmatrix} g_{11} & g_{12} & g_{13} & \cdots & g_{1n} \\ g_{21} & g_{22} & g_{23} & \cdots & g_{2n} \\ g_{31} & g_{32} & g_{33} & \cdots & g_{3n} \\ \vdots & \vdots & \vdots & \ddots & \vdots \\ g_{n1} & g_{n2} & g_{n3} & \cdots & g_{nn} \end{pmatrix}$$

Notice that this matrix is symmetric in the sense that each component above the diagonal is equal to one component which is its mirror-image below the diagonal (e.g., $g_{12} = g_{21}$). So all the components *above* the diagonal are reduplications of those *below* the diagonal, and the diagonal terms are not reduplicated. So we need only count how many terms there are on or below the diagonal. They form a triangle whose height is $n + 1$ units and whose base is n units. Thus there are $\frac{n(n+1)}{2}$ independent components, as Riemann claims, using the formula for the area of a triangle $A =$ (base $\times$ height)/2. The remaining components are therefore given by the total number, n^2, less this number. So the remaining components are $n^2 - \frac{n(n+1)}{2} = n\frac{n-1}{2}$. If we write dx^σ as a column vector, then ds^2 can be written in the neat form

$$ds^2 = \sum_{\tau\sigma} g_{\sigma\tau} dx^\sigma dx^\tau = \begin{pmatrix} g_{11} & g_{12} & g_{13} & \cdots & g_{1n} \\ g_{21} & g_{22} & g_{23} & \cdots & g_{2n} \\ g_{31} & g_{32} & g_{33} & \cdots & g_{3n} \\ \vdots & \vdots & \vdots & \ddots & \vdots \\ g_{n1} & g_{n2} & g_{n3} & \cdots & g_{nn} \end{pmatrix} \begin{pmatrix} dx_1 \\ dx_2 \\ dx_3 \\ \vdots \\ dx_n \end{pmatrix} \begin{pmatrix} dx_1 \\ dx_2 \\ dx_3 \\ \vdots \\ dx_n \end{pmatrix},$$

where matrix multiplication must be applied on the right-hand side.]

11. [In the Euclidean case, the distance function will take on the neat form $ds^2 = dx^2 + dy^2 + dz^2$ only if we choose the right variables; in polar coordinates the expression has a different look ($ds^2 = dr^2 + r^2 d\theta^2 + r^2 \sin^2\theta \, d\phi^2$) and in other curvilinear coordinates might be even more unrecognizably scrambled. That is, the *same* distance function might *look* very different in different coordinates: how can we distinguish real, physical effects from deceptive appearances that result from choosing different coordinates? Riemann knows already in the Euclidean case that one can always change variables back to Cartesian coordinates and recover the perspicuous form $ds^2 = dx^2 + dy^2 + dz^2$. The problem is how to do this in general, where one begins with an arbitrary coordinate system (not necessarily Cartesian!) and also the underlying geometry might truly not be Euclidean?

Taking the Euclidean plane as the paradigm of a *flat* manifold, Riemann requires that a flat manifold must be capable of having its distance function (line element) expressed as the square root of the sum of squares. To do this might require changing variables until such a form is found. All Riemann means here is that *if* variables can be found to do this, the manifold can be considered flat. Again, though, he realizes that he needs a way of finding out whether this can be done, facing some general line element in what may be peculiar coordinates.]

12. [To determine the curvature, Riemann uses *geodesics*, the shortest line between two given points. For instance, the geodesics on a sphere are the great circles. At each point in the manifold, he constructs an infinitesimal triangle centered about that point using geodesic lines for the sides of the triangle, also called "Riemannian normal coordinates." The geodesics give him a way of grasping something that lies behind mere arbitrary choice of coordinates, for the geodesics are not arbitrary. Riemann then examines whether the triangle is flat, by calculating the curvature of the plane in which it lies (the sum of its angles is $> 180°$ if it is convex or positively curved and $< 180°$ if concave or negatively curved). Because it is derived from geodesics (and not arbitrary lines drawn in arbitrary coordinates), the local curvature so derived does not depend on the choice of coordinates that may be used and thus is invariant. It is also important to note that in the infinitesmal vicinity of any point in a manifold there is no curvature; to lowest approximation the "tangent" to any point is a flat plane, and it is only as one moves further away from that point that one begins to notice curvature (which is why Riemann uses a triangle with one finite

side, x). So these continuous manifolds are always infinitesmally Euclidean; it is only as one considers their finite characteristics globally that the non-Euclidean properties come to light.]

13. [For Gauss's definition of curvature, see Gauss 2005, 9–21; the diagram and notes on 118 show how quantities like $(x_1 dx_2 - x_2 dx_1)$ emerge in the context of Gauss's definition.]

14. [In his "remarkable theorem" (*theorema egregium*), Gauss had shown that on a two-dimensional surface, the curvature he defined was invariant under bending of the surface without stretching; see Gauss 2005, v–vi, 20–21. Here Riemann points out a surprising general fact that he had drawn from Gauss's work: one can distinguish an *intrinsic* curvature from the *extrinsic* curvature. In his examples, a plane and a cylinder are both intrinsically flat, though extrinsically one is curved in the ambient three-dimensional space, and the other is not. That is, the Gaussian curvature of a plane or a cylinder are both everywhere zero; no surveying data could by themselves distinguish between them. On the other hand, a sphere has a constant, positive intrinsic curvature that cannot be "ironed out": we can print onto a flat page from a cylindrical drum but not from a sphere. The distinction between intrinsic and extrinsic curvature is confusing at first, but is important because it directs us to consider what can be said about a surface (or a manifold) from within itself, without confusing that with the separate issue of whether that surface is or is not embedded in still higher-dimensional space. The "two curvatures" referred to (in the case of the plane) are the curvatures along the two perpendicular directions that can be constructed at any point on the surface; one can bend a piece of paper independently in two orthogonal directions at any point, for instance. Gauss had proved that his curvature was proportional to the product of these two curvatures (Gauss 2005, 15), and did not depend on any extrinsic bending of the surface; Riemann asserts the same property for his curvature, which is the product of many more sub-curvatures along all possible orthogonal directions at the point in the manifold.]

15. [Note that here Riemann anticipates the arguments of Helmholtz and Lie that the existence and free mobility of rigid bodies implies that the curvature of space is constant. Unlike Helmholtz's initial opinion, Riemann does not assume that the curvature must be zero and also will examine in his final section the possibility that it is not constant.]

16. [Riemann does not explain this formula—the only one in the whole lecture—leaving us to conjecture what lies behind it. Weyl 1952, 57, 133–134, points out that from Riemann's assumptions (made more explicit in his Paris paper cited in the Introduction, note 14) it follows that, for a manifold with constant curvature α, an infinitesimal surface area on the curved manifold should be given by $\delta\sigma^2 = \frac{1}{4}\alpha\Delta x_{ij}\Delta x_{mn}$, where the infinitesimal coordinate area $\Delta x_{ij} = \delta x_i dx_j - dx_i \delta x_j$ (and the alternate notation for another, distinct infinitesimal length δx_i comes from Gauss 2005, 111–112, as diagrammed on 118); in terms of modern notation, $\alpha = R_{ijmn}$, now called the Riemann curvature tensor, which in this case is a constant. That is, the effect of the curvature measured by α

is to magnify or shrink the area indicated on the coordinate axes by a factor of $\frac{1}{4}\alpha$ (the factor $\frac{1}{4}$ appears because of the symmetries of the curvature tensor such as $R_{ijmn} = R_{mnij}$ reduce the number of its independent components; see note 13 above and Weyl 1952, 57). It follows that one could compensate for this constant magnification or shrinkage by introducing new coordinates such that the distance factor r is modified so that $r^*/r = 1 + \frac{\alpha}{4}r{*}^2$. Because $r^2 = \sum x_\alpha^2$, the individual coordinates x_i must each be modified by the same factor $x_i^* = x_i/(1+\frac{\alpha}{4}r^{*2})$ with the result that the line element $ds^2 = \sum dx_i^{*2}$ would be given by the form given here by Riemann. Gray 2005a, 510, gives the helpful example of a sphere, on which one could choose to measure distance according to the formula $ds = \sqrt{dx^2 + dy^2}/[1 + (\frac{1}{4r^2})(x^2 + y^2)]$, which corresponds to saying that the point (x, y) is specified in a plane tangent to the North Pole such that we are measuring the distance to the South Pole along a straight line (a great circle on the earth). In this case, the Gaussian curvature is $1/r^2$, agreeing with Riemann's definitions and his factor of $\frac{1}{4}$. For further discussion of this passage, see also Weyl's commentary in Riemann 1990, 758–766. Rosenfeld 1988, 286–287, 295, points out that this formula generalizes Ferdinand Minding's line element (1839) for a surface of constant curvature in terms of its Gaussian curvature.]

17. [Though Riemann does not mention it expressly, the case of negative curvature corresponds to Lobachevsky's "imaginary geometry," to hyperbolic or pseudo-spherical surfaces. As Lobachevsky showed, the angle of parallelism depends on the distance between the lines. As a consequence, in his geometry there is a largest possible triangle, determined by the behavior of the angle of parallelism.]

18. [Here Riemann anticipates Einstein's notion that it might be possible for the universe to be unbounded and yet finite, if its curvature were essentially that of a sphere; see below, 151–152.]

19. [In contrast, G. W. F. Hegel had argued that space as "self-externality ... is therefore absolutely continuous"; see Rosenfeld 1988, 198–199. Riemann's last thoughts are his most provocative, even today. By directing our attention to the possibility that solid bodies and rays of light "lose their validity in the infinitely small," he indicates the way in which the presumed continuity of space might finally be grounded on something not continuous. It is purely an assumption, he notes, that bodies exist independently of position and hence can be of any size (as is not true in Lobachevsky's geometry, for instance). Since this is not necessary, only empirical observation can really tell us whether or not there is some fundamental lower limit to the size of such bodies or light rays. Riemann suggests that it could not emerge from a geometrical continuum, because endless divisibility lacks any inner self-determination for distance relations. See the Introduction regarding the unresolved implications of this.]

20. [Though what Riemann might mean by such "binding forces" is tantalizingly unclear, see the following fragments from his notebooks for some clues as to the direction of his thought.]

Two Excerpts from Riemann's *Nachlass* (1853)

Bernhard Riemann

New mathematical principles of natural philosophy

Although the heading of this essay will hardly arouse a favorable expectation in most readers, yet it appeared to me to express best its overall drift.[1] Its objective is to penetrate into the depths of nature beyond the foundations of astronomy and physics laid down by Galileo and Newton. Admittedly, this speculation can have no immediate practical use for astronomy, but I hope that this circumstance will not, in the eyes of the reader of this page, undermine their interest ...

The foundation of the general laws of motion of ponderable bodies set out at the beginning of Newton's *Principia* lies in their inner nature (*Zustand*). Let us try to conclude something about this foundation, proceeding by analogy with our own inner perceptions. New masses of representations (*Vorstellungsmassen*) come before us incessantly and very rapidly disappear from our consciousness. We observe a constant activity of our soul. Underlying every act of the soul, something persists in its foundation, which shows itself as such on special occasions (through remembering), though it has no permanent impact on the phenomena. Thus, through every thought-act, something persistent incessantly comes before our soul, which, however, does not have a permanent impact on the world of phenomena. Underlying every action of our soul, there is something permanent that comes into the soul through this act, but which at the same moment entirely disappears from the world of phenomena.[2]

On account of this fact, I propose the hypothesis that the universe is filled with a substance (*Stoff*) that constantly flows into the ponderable atoms, where it disappears from the world of appearances (the corporeal world).[3]

The two hypotheses can be replaced by one hypothesis that in all ponderable atoms, substance from the corporeal world constantly enters into the spiritual world (*Geisteswelt*). The cause of the disappearance of the substance there must be the spiritual substance having been fashioned there directly beforehand. According to this, ponderable bodies are the place where the spiritual world intervenes in the corporeal world.*

*Into every ponderable atom at every moment a certain quantity of substance enters, proportional to the force of gravitation, and then disappears.

This is a consequence of a psychology subscribing to Herbart's principles that substantiality does not belong to the soul, but to each singular representation fashioned in us.

First of all, this hypothesis will explain the force of universal gravitation, which, as is well known, is completely determined in every part of space if the potential function P of all ponderable masses is given in that part of space, or (what amounts to the same thing) if such a function of position P is given such that the ponderable mass in the interior of a closed surface S equals $\frac{1}{4\pi}\int \frac{\partial P}{\partial p}\, dS$.[4]

If one assumes that the space-filling substance is an incompressible homogenous fluid without inertia and that into each ponderable atom in proportion to its mass there constantly flow equal quantities [of the substance] in equal time intervals, the pressure the ponderable atom experiences will be (proportional to the velocity of the moving substance at the position of the atom (?)).

Thus, it is possible to express the force of universal gravitation on a ponderable atom in terms of the pressure of the space-filling substance in the atom's immediate vicinity and to think of the force of gravity as being dependent on this.

It follows necessarily from our hypothesis that the space-filling substance must propagate the waves that we perceive as light and heat. [...]

Gravitation and light

The Newtonian explanation of falling motion and the motions of the heavenly bodies proceeds from the assumption of the following causes (*Ursachen*):

(1) There exists an infinite space with the properties geometry ascribes to it, in which ponderable bodies change their positions only by continuous motion.

(2) At every ponderable point, there exists at every moment a cause (*Ursache*) determined in magnitude and direction, by virtue of which the point has a certain motion (matter in determined states of motion). The measure of this cause is the velocity.†

The appearances to be explained here do not yet lead to the assumption of various masses for ponderable bodies.

(3) At every point of space exists at every moment a cause determined in magnitude and direction (accelerating force), which communicates a certain motion to every ponderable point existing there, indeed it communicates the same motion to all points, motion that is then compounded geometrically with the motion the point already had.

(4) At every ponderable point, there exists a quantitatively determined cause (absolute force of gravity), by virtue of which at each point in space there is an accelerating force inversely proportional to the square of the distance from that ponderable point and directly proportional to the point's force of gravity. This accelerating force is compounded geometrically with all the other accelerating forces occurring there.‡

---

†If a material body were alone in space, it would not change its location or would move through space along straight lines with unchanged velocity. This law of motion cannot be explained by the principle of sufficient reason. That the body continues its motion must have a cause that can only lie in the inner state (*Zustand*) of matter.

‡At two different places, the same ponderable point would exhibit changes in its motion, so that the direction of the motion coincides with the direction of the forces and its magnitude is in the ratio of the forces.

I seek the cause determined in magnitude and direction (accelerating force of gravity) at every point of space according to (3) in the state of motion of the continuous substance spread throughout the entire infinite space. In particular, I assume that the direction of this motion is equal to the direction of the force inferred from it and that its velocity is proportional to the magnitude of the force. Thus, this substance may be thought of as a physical space whose points move in geometrical space.[5]

On this assumption, all effects proceeding from ponderable bodies and impacting ponderable bodies via empty space must be propagated through this substance. Therefore, the forms of motion of light and heat the heavenly bodies send forth to each other must also be forms of motion of this substance. These two phenomena, gravitation and light traveling through empty space, must be the same, but they are the only ones that must be explained *purely* as motions of this substance.[6]

I now assume that the real motion of the substance in empty space is compounded of the motion that must be assumed for the explanation of gravitation and the motion that must be assumed for the explanation of light.

The further development of this hypothesis falls into two parts, insofar as we seek

1. The laws of motion of the substance that must be assumed for the explanation of phenomena.
2. The causes through which these motions can be explained.

The first of these tasks is mathematical, the second metaphysical. In regard to the latter, I remark in advance that its goal is not an explanation through causes that tend to alter the distance between two points in the substance. This method of explanation through forces of attraction and repulsion owes its general employment in physics not to immediate evidence (particularly in accordance with rationality), nor, apart from the case of electricity and gravity, to its particular facility, but rather much more to the circumstance that the Newtonian law of attraction, contrary to the intention of its discoverer, has for so long been taken as not capable of any further explanation.[§]

The force, divided by the change of motion, therefore yields the exact same quotient for the same ponderable point. This quotient is different for different ponderable points and is called their masses.

[§]Newton says: "That gravity should be innate, inherent, and essential to matter, so that one body may act upon another at a distance through a vacuum, without the mediation of anything else, by and through which their action and force may be conveyed from one to another, is to me so great an absurdity, that I believe no man who has in philosophical matters a competent faculty of thinking can ever fall into it." See the third letter to Bentley. [original in English]

Notes

[The German text of these fragments from Riemann's *Nachlass* comes from Riemann 1990, 560–561, 564–566; the manuscript bears the notation "found March 1, 1853." For a further account of Riemann's *Nachlass*, see Neuenschwander 1988.]

1. [Though his letters (and his opening paragraph) indicate that he later planned to publish them, Riemann clearly was still working on these writings (note his own inserted (?) below, indicating his awareness of the questionability of his hypothesis). He also may have wanted not to arouse the ire of powerful Kantians like Lotze. Riemann's title directly echoes Newton's *magnum opus*, his *Mathematical Principles of Natural Philosophy*. See, in particular, Laugwitz 1999, 281–287.]

2. [Here Riemann seems clearly inspired by Herbart's psychology, whom he echoes in this description of the psyche's actuality residing not in any substantial soul but in the mass of representations and experience that constitute our experience, as Riemann says in his footnote (and is discussed in the Introduction).]

3. [At first glance, Riemann's *Stoff* seems to a kind of cosmic ether or universal fluid medium of the sort that Descartes already had put forward, yet that is belied by Riemann's surprising assertion that it disappears from the material world into the spiritual (*Geistlich*), with the important ambiguity that the German *Geist* can suggest both mind as well as spirit. The context of Herbartian psychology may closely inform Riemann's use of *Geist* here as an understanding of mind as well as soul. For the earlier history of the concept of *Geist* in German natural philosophy, see Wise 1981, 272–275.]

4. [Here Riemann is making use what in electromagnetism is now generally called Gauss's Law: the integral of the electric flux over a closed surface equals 4π times the net charge enclosed by that surface: $\oiint \vec{E} \cdot \vec{dS} = 4\pi q_{enclosed}$. The same law holds true for any inverse square law force, like Newtonian universal gravitation. If P is the Newtonian gravitational potential and p the position variable, then the gravitational field is given by $\frac{\partial P}{\partial p}$ so that Riemann's integral is the gravitational equivalent of $\oiint \vec{E} \cdot \vec{dS}$ and hence equals the enclosed mass over 4π.]

5. [This assertion may set Riemann's hypotheses apart from, say, Newton's own attempts to ground universal gravitation on the behavior of some sort of medium, for Riemann here seems to point to the possibilitity that space itself might behave "physically," as compared with purely geometrical space. One wonders how close this finally may be to Einstein's developed conception of space-time geometry (determined by mass-energy distribution) as the physical cause of gravitation.]

6. [Here too Riemann anticipates Einstein's arguments that both gravitation and light must travel at the same speed, though in general relativity that speed is no longer the simple constant $c = 3 \times 10^8$ m/sec assumed in special relativity.

To what extent did Riemann also realize that this propagation of gravitational effect would necessitate gravitational waves, fully analogous to light waves? His attention is focused on unifying light with gravitation, it is unclear to what extent he saw a necessary connection between electricity and magnetism.]

On the Factual Foundations of Geometry (1866)

Hermann von Helmholtz

Investigations into how localization in the visual field comes to pass have led the author also to reflect on the origins of spatial intuition in general.[1] This leads first of all to a question whose answer definitely belongs to the sphere of exact sciences, namely, which propositions of geometry express truths of factual significance and which, on the contrary, are only definitions or consequences of definitions and their particular manner of expression? This investigation is completely independent of the further question: where does our knowledge about propositions of factual meaning come from? The former question is not quite so easily answered, as is so often done, because the spatial figures (***Raumgebilde***) of geometry are ideals that corporeal structures of the actual world can only approximate, without ever entirely satisfying the claim of the concept, and because in testing rigid bodies for the invariance of their form, the correctness of their planes and straight lines, in fact we must use the very same geometrical propositions we sought to prove.[2]

On the other hand, we can easily convince ourselves that (as the further course of this paper will show) the series of geometric axioms commonly put forward in elementary geometry is unsatisfactory; indeed, another sequence of further facts is here tacitly assumed. To be sure, newer textbooks seek to complement Euclid's axioms, but without a principle through which we could know whether this completion is itself complete. Because we can form an intuitive idea only of those spatial relations that can conceivably be exhibited in real space, we are misled by this intuitiveness to take something for granted as obvious that in truth is only a special and not at all obvious characteristic of the external world we find before us.[3]

Analytic geometry relieves us from this difficulty by using pure concepts of magnitude and relies on no intuition for its proofs. Therefore, to answer the above-mentioned question, one could follow this direction and find out which analytical characteristics of space and spatial magnitudes must be presupposed in order to ground the propositions of analytic geometry completely from the beginning.

The author had already begun such an investigation and had completed it in the main when Riemann's habilitation lecture "On the Hypotheses That Lie at the Foundations of Geometry" was made public, in which an identical investigation is carried out, having only a slightly different formulation of the question. On this occasion, we learned that Gauss had also worked on the same subject matter, of which his famous essay on the curvature of surfaces is the only published part of that investigation.[4]

Riemann begins by explaining that the general characteristics of space, its continuity, its multiplicity of dimensions, can be analytically expressed in such a fashion that every particular individual in the manifold that space manifests, that is, every point, can be determined through the measurement of n magnitudes, each a continuous and independently variable (coordinate). When n such magnitudes are needed, the space is what he calls an n-fold extended manifold and we ascribe to it n dimensions.

The system of colors is also a similar threefold-extended manifold.[5]

Now every line element in space is comparable in magnitude with any other, as far as can be judged. If u, v, w are measurements of any kind whatever that determine the location of a point and $u + du, v + dv, w + dw$ those of a nearby point, it follows that the magnitude of the line element ds in our actual space always equals the square root of a homogenous second-degree function of the magnitudes du, dv, dw, whatever the nature of the measurements u, v, w. We can call this proposition the most general form of the Pythagorean Theorem, which constitutes the central point of the entire investigation. It has a high degree of generality, since it is entirely independent of any particular system of measurement.

Riemann posits this expression for the line element as a hypothesis by showing that it is the simplest algebraic form that accords with the requirements of the problem. But he emphatically acknowledges its hypothetical character and mentions the possibility that ds could be regarded as the fourth root of a homogenous fourth-degree function of du, dv, and dw.

The further course of Riemann's investigation will be most intuitively captured if we confine ourselves to the discussion of two dimensions. In that case, it already follows from Gauss's investigation of curved surfaces that the most general form of a space of two dimensions, where the above-mentioned most general form of the Pythagorean Theorem is valid for the line element, is any curved surface whatever in our actual space in which the spatial determinations are made according to the ordinary rules of analytic geometry.

If figures of finite magnitude are moveable along all parts of such a surface without changing their relative measurements on the surface itself and if these figures can be rotated about any point whatever, it must be the case that the surface has constant curvature, that is, it must be a spherical surface or must be generated from such a surface by bending without stretching.[6]

If the extension of such a surface is infinite, it must be a plane or must be generated from a plane through bending without stretching.

Riemann now expands these propositions to include any number of dimensions and shows how to determine the curvature in this case. According to this investigation, the most general form of a space of three dimensions is a spatial construct delimited by any three equations within a space of six dimensions.[7]

Having solved the general problem, he qualifies the solution by adding the requirement that finite spatial constructs should be movable everywhere and rotatable in any direction without changing their form. In that case, the curvature of such an imaginary space must be constant, and if such a space is infinitely extended, the coefficient must equal zero. In this last case, such a space has the same attributes as our actual space and can be characterized as *flat* in relation to imaginary spaces of higher dimension.

My own investigation and its results are, for the most part, implicitly contained in Riemann's work. Only in one respect does it add something new concerning the proof of the generalized Pythagorean Theorem, which Riemann took as the starting point of his investigation. For the requirement that Riemann introduces only at the conclusion of his investigation, namely that spatial constructs exhibit the same degree of mobility without change of form that is assumed in geometry, I had introduced from the very start. This requirement so restricts the possibilities for hypotheses one can make about the expression of the line element that only the one Riemann adopted remains, excluding all other hypotheses.

My point of departure was that all original spatial measurement depends on asserting congruence and that, therefore, the system of spatial measurement must presuppose the same conditions on which alone it is meaningful to assert congruence.

The assumptions of my investigation are:

1) *Continuity and dimensions.* In spaces of n dimensions, the location of each point is determinable through measurement of n continuous and independent variables, so that (with the possible exception of certain points, lines, surfaces, or, speaking more generally, fixed structures of less than n dimensions) for each motion of the point, these variables, which serve as coordinates, undergo continuous change and at least one of them remains not unchanged.

2) *The existence of moving and rigid bodies.* Among the $2n$ coordinates of each pair of points of a rigid body that is moved, an equation holds that is the same for all congruent point pairs.

Though nothing further about the specific character of this equation will be said here, it is tightly constrained because for m points there are $m(m-1)/2$ equations in which mn unknown magnitudes are included, of which, however, only $n(n-1)/2$ have to remain arbitrarily changeable, in consequence of the next postulate. For if m is greater than $(n+1)$, there are more equations than unknowns and since all these equations must be constructed in an analogous fashion, this condition can only be satisfied by specific types of equations.[8]

3) *Free mobility.* Each point can pass over into any other along a continuous path. For the various points of one and the same rigid system exist only the constraints of motion that are contingent on the equations connecting the coordinates of each pair of points.

From 2) and 3) it follows that when a rigid system of points A in a certain position can be brought into congruence with a second system B, the same could take place also in any other position of A. For in the same way that A is made to pass over into the second position, B can be made to pass over there as well.

4) *The invariance of the form of rigid bodies under rotation.* If a body is moved so that $n - 1$ of its points remain unmoved and these are selected so that every other point of the body can only travel along a line, then continuous rotation without inversion will lead back to the initial position.

This last proposition which, as my research shows, is not implied by the preceding propositions, accords with the property of functions of complex variables called *mondromy.*[9]

As soon as these four conditions are met, it follows from pure analysis that a homogenous function of the second degree of the magnitudes du, dv, dw exists that is left invariant under rotation, thus yielding the magnitude of the line element independent of direction.*

With this, Riemann's point of departure has been reached and, going further along his path, it follows that when the number of dimensions is fixed at three and the infinite extension of space is assumed, no other geometry is possible beside Euclid's.

The first postulate, which Riemann also set out, is nothing other than the analytic definition of the concepts of the continuity of space and its manifold extension.

Postulates 2 to 4 must evidently be presupposed if congruence is to be meaningful at all. Hence, these presuppositions are the conditions for the possibility of congruence and, though often not clearly stated, constitute the foundation for the basic proofs of geometry, which founds all spatial measurement on congruence.

The system of these postulates thus makes no assumption that the ordinary form of geometry does not also make; it is complete and sufficient even without the specific axioms about the existence of straight lines and planes and without the parallel postulate. Theoretically speaking, it has the advantage that its completeness can be more easily checked.

It is noteworthy that in this case it is more clearly revealed how a certain character of rigidity and a particular degree of mobility of natural bodies is presupposed in order that such a system of measurement as that given in geometry can have factual significance at all. The independence of the congruence of rigid point-systems from place, location, and the system's relative rotation is the fact on which geometry is grounded.

This becomes even clearer when we compare space with other multiply extended manifolds, for example the system of colors. In this case, as long as we have no other method of measurement than through the law of color mixing, there exists, unlike in space, no relation of magnitudes between any two points that can be compared with that between two other points. Instead, there exists a relation between groups of any three points that also must lie in a straight line (that is, in groups of any three colors, among which any one is mixable from the other two).

---

*The detailed mathematical proof will appear in the Proceedings of the Royal Göttingen Society.[10]

We find another difference in the field of vision of a single eye, where no rotations are possible so long as we confine ourselves to natural eye movements. In my *Handbook of Physiological Optics* and in an earlier lecture given here (May 5, 1865), I have discussed which characteristic changes result from such measurements undertaken by the eye.[11]

As with every physical measurement, spatial measurement must depend on the invariable law of uniformity in natural phenomena.

(Additional note 1868) The essay referred to[10] contains an outline of my own researches that proves that if we desire to find the degree of rigidity and mobility of natural bodies attributable to our space in a space of otherwise unknown properties, the square of the line element ds would have to be a homogenous second-degree function of infinitely small increments of the arbitrarily chosen coordinates u, v, w. This proposition is there characterized as the most general form of the Pythagorean Theorem. The proof of this proposition vindicates the assumption of Riemann's investigations into space. I have found nothing to change in this part of my work.

In addition, in that essay I have given a brief list of further consequences following from Riemann's investigations, based on an as-yet unpublished and incompletely worked-out part of my researches, in which an error has crept in, because at that time I had not realized that a certain constant that I had believed must be taken to be real is still meaningful when taken as imaginary. Thus the assertion made in that essay that space, if supposed to be infinitely extended, must be flat (in Riemann's sense) is false.

This is the result, in particular, of the highly interesting and important investigations of Beltrami,[12] in which he explored the theory of surfaces and spaces of constant negative curvature and showed their agreement with the imaginary geometry that Lobachevsky had earlier presented. In this geometry, space is infinitely extended in all directions; figures congruent to a given direction can be constructed in all parts of this space; between any two points only one shortest line is possible, but the parallel postulate does not hold.

Notes

[The original text, "Ueber die thatsächlichen Grundlagen der Geometrie," is in Helmholtz 1883, 620–617. For Helmholtz's longer 1868 account of this work, see note 8 below. Note that the term *thatsächlichen* suggests both "actual" as well as "factual."]

1. [For Helmholtz's investigations on visual localization, see his *Handbook of Physiological Optics* (Helmholtz 1962a, 3:1–37, 228–231, 560–577); he also returns to this question in the next paper in this anthology, "On the Origins and Meaning of Geometrical Axioms."]

2. [Helmholtz uses the phrase "feste Körper" to mean "rigid bodies," as I will translate it; later, Einstein will use this phrase slightly differently and I will there translate it as "solid bodies"; see 162 below.]

3. [Unlike those who wanted to supplement Euclid by adding axioms (of completeness, for example) to address lacunae they perceived in his proofs, Helmholtz's main concern is to show that the ultimate foundations of geometry are facts, not merely hypotheses or axioms.]

4. [Though delivered in 1854, Riemann's lecture was only published in 1868; for Gauss's writings on curved surfaces, see Gauss 2005.]

5. [See 34, note 4, above.]

6. [Helmholtz himself will in part correct this assertion at the end of his paper by adding pseudo-spherical (Lobachevskian) surfaces, not just spherical.]

7. [For the counting of six dimensions, see below, 115, note 3.]

8. [For the counting of the equations, see above, 37, note 10.]

9. [*Monodromy* means the invariance of a figure upon a rotation of 360°.]

10. [The essay in question is Helmholtz 1868. For a translation of this longer account of Helmholtz's arguments, "On the facts underlying geometry," see Helmholtz 1977, 39–71; the reader should note, however, that this translation renders the crucial term "rigid body" (*fester Körper*) by "fixed body," which is misleading because it suggests that the body is "fixed" in the sense of stationary or unmoving, rather than the correct sense of rigid, having sides of invariant length. Sophus Lie argued that Helmholtz's mathematical argument contained essential faults; see his paper "Sur les fondements de la Géométrie" (1892), in Lie 1935, 477–479, and (more fully) Lie 1967. Because of this, the whole topic is generally called the Helmholtz-Lie space problem. There is an excellent and concise modern account of Helmholtz's argument in Adler, Bazin, and Schiffer 1975, 10–16; see also Stein 1977, 36–39, Torretti 1978, 155–179, Scholz 1980, 113–123, Laugwitz 1999, 239–240, and Tazzioli 2003.]

11. [Helmholtz seems to be referring to his 1865 paper "Ueber stereoskopisches Sehen," Helmholtz 1883, 2:492–496; for references in his *Treatise*, see Helmholtz 1962a, 3:281–368.]

12. [Helmholtz cites Beltrami 1868a, 1868b; for a description of Beltrami's work, see Bonola 1955, 127–139, Scholz 1980, 101–113, and Helmholtz's following essay in this anthology.]

The Origin and Meaning of Geometrical Axioms (1870)

Hermann von Helmholtz

My object in this article is to discuss the philosophical bearing of recent inquiries concerning geometrical axioms and the possibility of working out analytically other systems of geometry with other axioms than Euclid's. The original works on the subject, addressed to experts only, are particularly abstruse, but I will try to make it plain even to those who are not mathematicians. It is of course no part of my plan to prove the new doctrines correct as mathematical conclusions. Such proof must be sought in the original works themselves.

Among the first elementary propositions of geometry, from which the student is led on by continuous chains of reasoning to the laws of more and more complex figures, are some which are held not to admit of proof, though sure to be granted by every one who understands their meaning. These are the so-called Axioms; for example, the proposition that if the shortest line drawn between two points is called straight there can be only one such straight line. Again, it is an axiom that through any three points in space, not lying in a straight line, a plane may be drawn, i.e., a surface which will wholly include every straight line joining any two of its points. Another axiom, about which there has been much discussion, affirms that through a point lying without a straight line only one straight line can be drawn parallel to the first; two straight lines that lie in the same plane and never meet, however far they may be produced, being called parallel. There are also axioms that determine the number of dimensions of space and its surfaces, lines and points, showing how they are continuous; as in the propositions, that a solid is bounded by a surface, a surface by a line and a line by a point, that the point is indivisible, that by the movement of a point a line is described, by that of a line a line or a surface, by that of a surface a surface or a solid, but by the movement of a solid a solid and nothing else is described.

Now what is the origin of such propositions, unquestionably true yet incapable of proof in a science where everything else is reasoned conclusion? Are they inherited from the divine source of our reason as the idealistic philosophers think,[1] or is it only that the ingenuity of mathematicians has hitherto not been penetrating enough to find the proof? Every new votary, coming with fresh zeal to geometry, naturally strives to succeed where all before him have failed. And it is quite right that each should make the trial afresh; for, as the question has hitherto stood, it is only by the fruitlessness of ones own efforts that one can be convinced of the impossibility of finding a proof. Meanwhile solitary inquirers are always from time to time appearing who become so deeply entangled in complicated trains of reasoning that they can no longer discover their mistakes and believe they have solved the problem. The axiom of parallels especially has called forth a great number of seeming demonstrations.

The main difficulty in these inquiries is and always has been the readiness with which results of everyday experience become mixed up as apparent necessities of thought with the logical processes, so long as Euclid's method of constructive intuition is exclusively followed in geometry. In particular it is extremely difficult, on this method, to be quite sure that in the steps prescribed for the demonstration we have not involuntarily and unconsciously drawn in some most general results of experience, which the power of executing certain parts of the operation has already taught us practically. In drawing any subsidiary line for the sake of his demonstration, the well-trained geometer asks always if it is possible to draw such a line. It is notorious that problems of construction play an essential part in the system of geometry. At first sight, these appear to be practical operations, introduced for the training of learners; but in reality they have the force of existential propositions. They declare that points, straight lines or circles, such as the problem requires to be constructed, are possible under all conditions, or they determine any exceptions that there may be. The point on which the investigations turn that we are going to consider is essentially of this nature. The foundation of all proof by Euclid's method consists in establishing the congruence of lines, angles, plane figures, solids, &c. To make the congruence evident, the geometrical figures are supposed to be applied to one another, of course without changing their form and dimensions. That this is in fact possible we have all experienced from our earliest youth. But, when we would build necessities of thought upon this assumption of the free translation of fixed figures with unchanged form to every part of space, we must see whether the assumption does not involve some presupposition of which no logical proof is given. We shall see later on that it does contain one of most serious import. But if so, every proof by congruence rests upon a fact which is obtained from experience only.

I offer these remarks at first only to show what difficulties attend the complete analysis of the presuppositions we make in employing the common constructive method. We evade them when we apply to the investigation of principles the analytical method of modern algebraical geometry. The whole process of algebraical calculation is a purely logical operation; it can yield no relation between the quantities submitted to it that is not already contained in the equations which give occasion for its being applied. The recent investigations have accordingly been conducted almost exclusively by means of the purely abstract methods of analytical geometry.

However, after discovering by the abstract method what are the points in question, we shall best get a distinct view of them by taking a region of narrower limits than our own world of space. Let us, as we logically may, suppose reasoning beings of only two dimensions to live and move on the surface of some solid body.[2] We will assume that they have not the power of perceiving anything outside this surface, but that upon it they have perceptions similar to ours. If such beings worked out a geometry, they would of course assign only two dimensions to their space. They would ascertain that a point in moving describes a line, and that a line in moving describes a surface. But they could as little represent to themselves what, further spatial construction would be generated by a surface moving out of itself, as we can represent what would be generated by a solid moving out of the space we know.

By the much abused expression "to represent" or "to be able to think how something happens" I understand—and I do not see how anything else can be understood by it without loss of all meaning — the power of imagining the whole series of sensible impressions that would be had in such a case. Now as no sensible impression is known relating to such an unheard-of event as the movement to a fourth dimension would be to us, or as a movement to our third dimension would be to the inhabitants of a surface, such a "representation" is as impossible as the "representation" of colours would be to one born blind, though a description of them in general terms might be given to him. Our surface-beings would also be able to draw shortest lines in their superficial space. These would not necessarily be straight lines in our sense, but what are technically called *geodetic* lines of the surface on which they live, lines such as are described by a tense thread laid along the surface and which can slide upon it freely.[3] I will henceforth speak of such lines as the *straightest* lines of any particular surface or given space, so as to bring out their analogy with the straight line in a plane.

Now if beings of this kind lived on an infinite plane, their geometry would be exactly the same as our planimetry.[4] They would affirm that only one straight line is possible between two points, that through a third point lying without this line only one line can be drawn parallel to it, that the ends of a straight line never meet though it is produced to infinity, and so on. Their space might be infinitely extended, but even if there were limits to their movement and perception, they would be able to represent to themselves a continuation beyond these limits, and thus their space would appear to them infinitely extended, just as ours does to us, although our bodies cannot leave the earth and our sight only reaches as far as the visible fixed stars.

But intelligent beings of the kind supposed might also live on the surface of a sphere. Their shortest or straightest line between two points would then be an arc of the great circle passing through them. Every great circle passing through two points is by these divided into two parts, and if they are unequal the shorter is certainly the shortest line on the sphere between the two points, but also the other or larger arc of the same great circle is a geodetic or straightest line, i.e., every smaller part of it is the shortest line between its ends. Thus the notion of the geodetic or straightest line is not quite identical with that of the shortest line. If the two given points are the ends of a diameter of the sphere every plane passing through this diameter cuts semicircles on the surface of the sphere all of which are shortest lines between the ends; in which case there is an infinite number of equal shortest lines between the given points. Accordingly, the axiom of there being only one shortest line between two points would not hold without a certain exception for the dwellers on a sphere.

Of parallel lines the sphere-dwellers would know nothing. They would declare that any two straightest lines, sufficiently produced, must finally cut not in one only but in two points. The sum of the angles of a triangle would be always greater than two right angles, increasing as the surface of the triangle grew greater. They could thus have no conception of geometrical similarity between greater and smaller figures of the same kind, for with them a greater triangle must have different angles from a smaller one. Their space would be unlimited, but would be found to be finite or at least represented as such.

It is clear, then, that such beings must set up a very different system of geometrical axioms from that of the inhabitants of a plane or from ours with our space of three dimensions, though the logical powers of all were the same; nor are more examples necessary to show that geometrical axioms must vary according to the kind of space inhabited. But let us proceed still farther.

Let us think of reasoning beings existing on the surface of an egg-shaped body. Shortest lines could be drawn between three points of such a surface and a triangle constructed. But if the attempt were made to construct congruent triangles at different parts of the surface, it would be found that two triangles with three pairs of equal sides would not have their angles equal. The sum of the angles of a triangle drawn at the sharper pole of the body would depart farther from two right angles than if the triangle were drawn at the blunter pole or at the equator. Hence it appears that not even such a simple figure as a triangle can be moved on such a surface without change of form. It would also be found that if circles of equal radii were constructed at different parts of such a surface (the length of the radii being always measured by shortest lines along the surface) the periphery would be greater at the blunter than at the sharper end.

We see accordingly that, if a surface admits of the figures lying on it being freely moved without change of any of their lines and angles as measured along it, the property is a special one and does not belong to every kind of surface. The condition under which a surface possesses this important property was pointed out by Gauss in his celebrated treatise on the curvature of surfaces [Gauss 2005]. The "measure of curvature," as he called it, i.e., the reciprocal of the product of the greatest and least radii of curvature, must be everywhere equal over the whole extent of the surface.

Gauss showed at the same time that this measure of curvature is not changed if the surface is bent without distension or contraction of any part of it. Thus we can roll up a flat sheet of paper into the form of a cylinder or of a cone without any change in the dimensions of the figures taken along the surface of the sheet. Or the hemispherical fundus of a bladder[5] may be rolled into a spindle-shape without altering the dimensions on the surface. Geometry on a plane will therefore be the same as on a cylindrical surface; only in the latter case we must imagine that any number of layers of this surface, like the layers of a rolled sheet of paper, lie one upon another and that after each entire revolution round the cylinder a new layer is reached.

These observations are meant to give the reader a notion of a kind of surface the geometry of which is on the whole similar to that of the plane, but in which the axiom of parallels does not hold good, namely, a kind of curved surface which geometrically is, as it were, the counterpart of a sphere, and which has therefore been called the *pseudospherical surface* by the distinguished Italian mathematician, E. Beltrami, who has investigated its properties [Beltrami 1868a, 1868b].[6] It is a saddle-shaped surface of which only limited pieces or strips can be connectedly represented in our space, but which may yet be thought of as infinitely continued in all directions, since each piece lying at the limit of the part constructed can be conceived as drawn back to the middle of it and then continued. The piece displaced must in the process change its flexure but not its dimensions, just as happens with a sheet of paper moved about a cone formed out of a plane rolled up. Such a sheet fits the conical surface in every part, but must be more bent near the vertex and cannot be so moved over the vertex as to be at the same time adapted to the existing cone and to its imaginary continuation beyond.

Like the plane and the sphere, pseudospherical surfaces have their measure of curvature constant, so that every piece of them can be exactly applied to every other piece, and therefore all figures constructed at one place on the surface can be transferred to any other place with perfect congruity of form and perfect equality of all dimensions lying in the surface itself. The measure of curvature as laid down by Gauss, which is positive for the sphere and zero for the plane, would have a constant negative value for pseudospherical surfaces, because the two principal curvatures of a saddle-shaped surface have their concavity turned opposite ways.

A strip of a pseudospherical surface may, for example, be represented by the inner surface (turned towards the axis) of a solid anchor-ring. If the plane figure *aabb* (Fig. 1) is made to revolve on its axis of symmetry *AB*, the two arcs *ab* will describe a pseudospherical concave-convex surface like that of the ring.

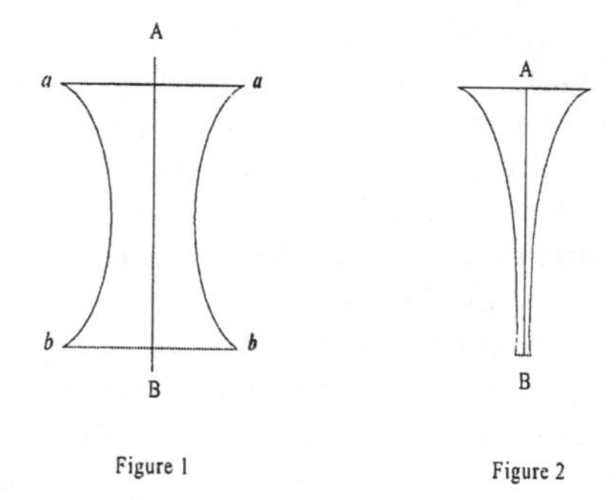

Figure 1 Figure 2

Above and below, towards *aa* and *bb*, the surface will turn outwards with ever-increasing flexure, till it becomes perpendicular to the axis and ends at the edge with one curvature infinite. Or, again, half of a pseudospherical surface may be rolled up into the shape of a champagne-glass (Fig. 2) with tapering stem infinitely prolonged. But the surface is always necessarily bounded by a sharp edge beyond which it cannot be directly continued. Only by supposing each single piece of the edge cut loose and drawn along the surface of the ring or glass, can it be brought to places of different flexure at which farther continuation of the piece is possible.

In this way too the straightest lines of the pseudospherical surface may be infinitely produced. They do not like those on a sphere return upon themselves, but, as on a plane, only one shortest line is possible between two given points. The axiom of parallels does not however hold good. If a straightest line is given on the surface and a point without it, a whole pencil of straightest lines may pass through the point, no one of which, though infinitely produced, cuts the first line; the pencil itself being limited by two straightest lines, one of which intersects one of the ends of the given line at an infinite distance, the other the other end.

As it happened, a system of geometry excluding the axiom of parallels was devised on Euclid's synthetic method, as far back as the year 1829, by N. I. Lobachevsky, professor of mathematics at Kazan,[7] and it was proved that this system could be carried out as consistently as Euclid's. It agrees exactly with the geometry of the pseudospherical surfaces worked out recently by Beltrami.

Thus we see that in the geometry of two dimensions a surface is marked out as a plane or a sphere or a pseudospherical surface by the assumption that any figure may be moved about in all directions without change of dimensions. The axiom that there is only one shortest line between any two points distinguishes the plane and the pseudospherical surface from the sphere, and the axiom of parallels marks off the plane from the pseudosphere. These three axioms are in fact necessary and sufficient to define as a plane the surface to which Euclid's planimetry has reference, as distinguished from all other modes of space in two dimensions.

The difference between plane and spherical geometry has been long evident, but the meaning of the axiom of parallels could not be understood till Gauss had developed the notion of surfaces flexible without dilatation and consequently that of the possibly infinite continuation of pseudospherical surfaces. Inhabiting a space of three dimensions and endowed with organs of sense for their perception, we can represent to ourselves the various cases in which beings on a surface might have to develop their perception of space; for we have only to limit our own perceptions to a narrower field. It is easy to think away perceptions that we have; but it is very difficult to imagine perceptions to which there is nothing analogous in our experience. When, therefore, we pass to space of three dimensions we are stopped in our power of representation by the structure of our organs and the experiences got through them which correspond only to the space in which we live.

There is however another way of treating geometry scientifically. All known space-relations are measurable, that is they may be brought to determination of magnitudes (lines, angles, surfaces, volumes). Problems in geometry can therefore be solved by finding methods of calculation for arriving at unknown magnitudes from known ones. This is done in analytical geometry, where all forms of space are treated only as quantities and determined by means of other quantities. Even the axioms themselves make reference to magnitudes. The straight line is defined as the *shortest* between two points, which is a determination of quantity. The axiom of parallels declares that if two straight lines in a plane do not intersect (are parallel), the alternate angles, or the corresponding angles, made by a third line intersecting them, are equal; or it may be laid down instead that the sum of the angles of any triangle is equal to two right angles. These are determinations of quantity.

Now we may start with this view of space, according to which the position of a point may be determined by measurements in relation to any given figure (system of co-ordinates), taken as fixed, and then inquire what are the special characteristics of our space as manifested in the measurements that have to be made, and how it differs from other extended quantities of like variety. This path was first entered by one too early lost to science, B. Riemann of Göttingen. It has the peculiar advantage that all its operations consist in pure calculation of quantities, which quite obviates the danger of habitual perceptions being taken for necessities of thought.

The number of measurements necessary to give the position of a point is equal to the number of dimensions of the space in question. In a line the distance from one fixed point is sufficient, that is to say, one quantity; in a surface the distances from two fixed points must be given; in space, the distances from three; or we require as on the earth longitude, latitude and height above the sea, or, as is usual in analytical geometry, the distances from three co-ordinate planes. Riemann calls a system of differences in which one thing can be determined by n measurements an "n-fold extended aggregate" or an "aggregate of n dimensions." Thus the space in which we live is a three-fold, a surface is a twofold and a line is a simple extended aggregate of points. Time also is an aggregate of one dimension. The system of colours is an aggregate of three dimensions, inasmuch as each colour, according to the investigations of Th. Young and Clerk Maxwell, may be represented as a mixture of three primary colours, taken in definite quantities. The particular mixtures can be actually made with the colour-top.[8]

In the same way we may consider the system of simple tones as an aggregate of two dimensions, if we distinguish only pitch and intensity and leave out of account differences of timbre. This generalisation of the idea is well-suited to bring out the distinction between space of three dimensions and other aggregates. We can, as we know from daily experience, compare the vertical distance of two points with the horizontal distance of two others, because we can apply a measure first to the one pair and then to the other. But we cannot compare the difference between two tones of equal pitch and different intensity with that between two tones of equal intensity and different pitch.[9] Riemann showed by considerations of this kind that the essential foundation of any system of geometry is the expression that it gives for the distance between two points lying in any direction from one another, beginning with the interval as infinitesimal.

He took from analytical geometry the most general form for this expression, that, namely, which leaves altogether open the kind of measurements by which the position of any point is given.* Then he showed that the kind of free mobility without change of form which belongs to bodies in our space can only exist when certain quantities yielded by the calculation†—quantities that coincide with Gausss measure of surface-curvature when they are expressed for surfaces—have everywhere an equal value. For this reason Riemann calls these quantities, when they have the same value in all directions for a particular spot, the measure of curvature of the space at this spot. To prevent misunderstanding I will once more observe that this so-called measure of space-curvature is a quantity obtained by purely analytical calculation and that its introduction involves no suggestion of relations that would have a meaning only for sense-perception. The name is merely taken, as a short expression for a complex relation, from the one case in which the quantity designated admits of sensible representation.

Now whenever the value of this measure of curvature in any space is everywhere zero, that space everywhere conforms to the axioms of Euclid; and it may be called a *flat* (*homaloid*) space in contradistinction to other spaces, analytically constructible, that may be called *curved* because their measure of curvature has a value other than zero. Analytical geometry may be as completely and consistently worked out for such spaces as ordinary geometry for our actually existing homaloid space.

If the measure of curvature is positive we have *spherical* space, in which straightest lines return upon themselves and there are no parallels. Such a space would, like the surface of a sphere, be unlimited but not infinitely great. A constant negative measure of curvature on the other hand gives *pseudospherical* space, in which straightest lines run out to infinity and a pencil of straightest lines may be drawn in any fattest surface through any point which do not intersect another given straightest line in that surface.

Beltrami [1868b] has rendered these last relations imaginable by showing that the points, lines and surfaces of a pseudospherical space of three dimensions can be so portrayed in the interior of a sphere in Euclid's homaloid space, that every straightest line or flattest surface of the pseudospherical space is represented by a straight line or a plane, respectively, in the sphere. The surface itself of the sphere corresponds to the infinitely distant points of the pseudospherical space; and the different parts of this space, as represented in the sphere, become smaller the nearer they lie to the spherical surface, diminishing more rapidly in the direction of the radii than in that perpendicular to them. Straight lines in the sphere which only intersect beyond its surface correspond to straightest lines of the pseudospherical space which never intersect.

---

*For the square of the distance of two infinitely near points the expression is a homogeneous quadric function of the differentials of their co-ordinates.

†They are algebraical expressions compounded from the co-efficients of the various terms in the expression for the square of the distance of two contiguous points and from their differential quotients.

Thus it appeared that space, considered as a region of measurable quantities, does not at all correspond with the most general conception of an aggregate of three dimensions, but involves also special conditions, depending on the perfectly free mobility of solid bodies without change of form to all parts of it and with all possible changes of direction, and, farther, on the special value of the measure of curvature which for our actual space equals, or at least is not distinguishable from, zero. This latter definition is given in the axioms of straight lines and parallels.

Whilst Riemann entered upon this new field from the side of the most general and fundamental questions of analytical geometry, I myself arrived at similar conclusions [Helmholtz 1868], partly from seeking to represent in space the system of colours, involving the comparison of one threefold extended aggregate with another, and partly from inquiries on the origin of our ocular measure for distances in the field of vision. Riemann starts by assuming the above-mentioned algebraical expression which represents in the most general form the distance between two infinitely, near points, and deduces therefrom the conditions of mobility of rigid figures. I, on the other hand, starting from the observed fact that the movement of rigid figures is possible in our space, with the degree of freedom that we know, deduce the necessity of the algebraic expression taken by Riemann as an axiom. The assumptions that I had to make as the basis of the calculation were the following.

First, to make algebraical treatment possible, it must be assumed that the position of any point A can be determined, in relation to certain given figures taken as fixed bases, by measurement of some kind of magnitudes, as lines, angles between lines, angles between surfaces and so forth. The measurements necessary for determining the position of A are known as its co-ordinates. In general the number of coordinates necessary to the complete determination of the position of a point marks the number of the dimensions of the space in question. It is further assumed that with the movement of the point A the magnitudes used as co-ordinates vary continuously.

Secondly, the definition of a solid body, or rigid system of points, must be made in such a way as to admit of magnitudes being compared by congruence. As we must not at this stage assume any special methods for the measurement of magnitudes, our definition can, in the first instance, run only as follows: Between the co-ordinates of any two points belonging to a solid body, there must be an equation which, however the body is moved, expresses a constant spatial relation (proving at last to be the distance) between the two points, and which is the same for congruent pairs of points, that is to say, such pairs as can be made successively to coincide in space with the same fixed pair of points.

However indeterminate in appearance, this definition involves most important consequences, because with increase in the number of points he number of equations increases much more quickly than the number of co-ordinates which they determine. Five points, A, B, C, D, E give ten different pairs of points $(AB, AC, AD, AE, BC, BD, BE, CD, CE, DE)$ and therefore ten equations, involving in space of three dimensions fifteen variable co-ordinates. But of these fifteen six must remain arbitrary if the system of five points is to admit of free movement and rotation, and thus the ten equations can determine only nine co-ordinates as functions of the six variables. With six points we obtain fifteen equations for twelve quantities, with seven points twenty-one equations for fifteen, and so on. Now from n independent equations we can determine n contained quantities, and if we have more than n equations, the superfluous ones must be deducible from the first n. Hence it follows that the equations which subsist between the co-ordinates of each pair of points of a solid body must have a special character, seeing that, when in space of three dimensions they are satisfied for nine pairs of points as formed out of any five points, the equation for the tenth pair follows by logical consequence. Thus our assumption for the definition of solidity becomes quite sufficient to determine the kind of equations holding between the co-ordinates of two points rigidly connected.

Thirdly, the calculation must further be based on the fact of a peculiar circumstance in the movement of solid bodies, a fact so familiar to us that but for this inquiry it might never have been thought of as something that need not be. When in our space of three dimensions two points of a solid body are kept fixed, its movements are limited to rotations round the straight line connecting them. If we turn it completely round once, it again occupies exactly the position it had at first.[10] This fact that rotation in one direction always brings a solid body back into its original position needs special mention. A system of geometry is possible without it. This is most easily seen in the geometry of a plane. Suppose that with every rotation of a plane figure its linear dimensions increased in proportion to the angle of rotation, the figure after one whole rotation through 360 degrees would no longer coincide with itself as it was originally. But any second figure that was congruent with the first in its original position might be made to coincide with it in its second position by being also turned through 360 degrees. A consistent system of geometry would be possible upon this supposition, which does not come under Riemann's formula.

On the other hand I have shown that the three assumptions taken together form a sufficient basis for the starting-point of Riemann's investigation, and thence for all his further results relating to the distinction of different spaces according to their measure of curvature.

It still remained to be seen whether the laws of motion as dependent on moving forces could also be consistently transferred to spherical or pseudospherical space. This investigation has been carried out by Professor Lipschitz of Bonn.[‡] It is found that the comprehensive expression for all the laws of dynamics, Hamilton's principle, may be directly transferred to spaces of which the measure of curvature is other than zero.[11] Accordingly, in this respect also the disparate systems of geometry lead to no contradiction.

We have now to seek an explanation of the special characteristics of our own flat space, since it appears that they are not implied in the general notion of an

[‡]Untersuchungen über die ganzen homogenen Functionen von n Differentialen (Borchardts *Journal für Mathematik*, Bde. lxx. 3, 71; lxxiiii. 3, 1); Untersuchung eines Problems der Variationsrechnung (Ibid. Bd. 1xxiv)

extended quantity of three dimensions and of the free mobility of bounded figures therein. Necessities of thought, involved in such a conception, they are not. Let us then examine the opposite assumption as to their origin being empirical, and see if they can be inferred from facts of experience and so established, or if, when tested by experience, they are perhaps to be rejected. If they are of empirical origin we must be able to represent to ourselves connected series of facts indicating a different value for the measure of curvature from that of Euclid's flat space. But if we can imagine such spaces of other sorts, it cannot be maintained that the axioms of geometry are necessary consequences of an a priori transcendental form of intuition, as Kant thought.

The distinction between spherical, pseudospherical and Euclid's geometry depends, as was above observed, on the value of a certain constant called by Riemann the measure of curvature of the space in question. The value must be zero for Euclid's axioms to hold good. If it were not zero, the sum of the angles of a large triangle would differ from that of the angles of a small one, being larger in spherical, smaller in pseudospherical space. Again, the geometrical similarity of large and small solids or figures is possible only in Euclid's space. All systems of practical mensuration that have been used for the angles of large rectilinear triangles, and especially all systems of astronomical measurement which make the parallax of the immeasurably distant fixed stars equal to zero (in pseudospherical space the parallax even of infinitely distant points would be positive), confirm empirically the axiom of parallels and show the measure of curvature of our space thus far to be indistinguishable from zero. It remains, however, a question, as Riemann observed, whether the result might not be different if we could use other than our limited baselines, the greatest of which is the major axis of the earths orbit.

Meanwhile, we must not forget that all geometrical measurements rest ultimately upon the principle of congruence. We measure the distance between points by applying to them the compass, rule or chain. We measure angles by bringing the divided circle or theodolite to the vertex of the angle. We also determine straight lines by the path of rays of light which in our experience is rectilinear; but that light travels in shortest lines as long as it continues in a medium of constant refraction would be equally true in space of a different measure of curvature. Thus all our geometrical measurements depend on our instruments being really, as we consider them, invariable in form, or at least on their undergoing no other than the small changes we know of as arising from variation of temperature or from gravity acting differently at different places.

In measuring we only employ the best and surest means we know of to determine what we otherwise are in the habit of making out by sight and touch or by pacing. Here our own body with its organs is the instrument we carry about in space. Now it is the hand, now the leg that serves for a compass, or the eye turning in all directions is our theodolite for measuring arcs and angles in the visual field.

Every comparative estimate of magnitudes or measurement of their spatial relations proceeds therefore upon a supposition as to the behaviour of certain physical things, either the human body or other instruments employed. The supposition may be in the highest degree probable and in closest harmony with all other physical relations known to us, but yet it passes beyond the scope of pure space-intuition.

It is in fact possible to imagine conditions for bodies apparently solid such that the measurements in Euclid's space become what they would be in spherical or pseudospherical space. Let me first remind the reader that if all the linear dimensions of other bodies and our own at the same time were diminished or increased in like proportion, as for instance to half or double their size, we should with our means of space-perception be utterly unaware of the change. This would also be the case if the distension or contraction were different in different directions, provided that our own body changed in the same manner and further that a body in rotating assumed at every moment, without suffering or exerting mechanical resistance, the amount of dilatation in its different dimensions corresponding to its position at the time. Think of the image of the world in a convex mirror. The common silvered globes set up in gardens give the essential features, only distorted by some optical irregularities. A well-made convex mirror of moderate aperture represents the objects in front of it as apparently solid and in fixed positions behind its surface. But the images of the distant horizon and of the sun in the sky lie behind the mirror at a limited distance, equal to its focal length. Between these and the surface of the mirror are found the images of all the other objects before it, but the images are diminished and flattened in proportion to the distance of their objects from the mirror. The flattening, or decrease in the third dimension, is relatively greater than the decrease of the surface-dimensions. Yet every straight line or every plane in the outer world is represented by a straight line or a plane in the image. The image of a man measuring with a rule a straight line from the mirror would contract more and more the farther he went, but with his shrunken rule the man in the image would count out exactly the same number of centimetres as the real man. And, in general, all geometrical measurements of lines or angles made with regularly varying images of real instruments would yield exactly the same results as in the outer world, all congruent bodies would coincide on being applied to one another in the mirror as in the outer world, all lines of sight in the outer world would be represented by straight lines of sight in the mirror. In short I do not see how men in the mirror are to discover that their bodies are not rigid solids and their experiences good examples of the correctness of Euclid's axioms. But if they could look out upon our world as we can look into theirs, without overstepping the boundary, they must declare it to be a picture in a spherical mirror, and would speak of us just as we speak of them; and if two inhabitants of the different worlds could communicate with one another, neither, so far as I can see, would be able to convince the other that he had the true, the other the distorted relations.[12] Indeed I cannot see that such a question would have any meaning at all so long as mechanical considerations are not mixed up with it.

Now Beltrami's representation of pseudospherical space in a sphere of Euclid's space is quite similar except that the background is not a plane as in the convex mirror, but the surface of a sphere, and that the proportion in which the images as they approach the spherical surface contract, has a different mathematical expression.[13] If we imagine then, conversely, that in the sphere, for the interior of which Euclid's axioms hold good, moving bodies contract as they depart from the centre like the images in a convex mirror, and in such a way that their representatives in pseudospherical space retain their dimensions unchanged,—observers whose bodies were regularly subjected to the same change would obtain the same results from the geometrical measurements they could make as if they lived in pseudospherical space.

We can even go a step further, and infer how the objects in a pseudospherical world, were it possible to enter one, would appear to an observer whose eye-measure and experiences of space had been gained like ours in Euclid's space. Such an observer would continue to look upon rays of light or the lines of vision as straight lines, such as are met with in flat space and as they really are in the spherical representation of pseudospherical space. The visual image of the objects in pseudospherical space would thus make the same impression upon him as if he were at the centre of Beltrami's sphere. He would think he saw the most remote objects round about him at a finite distance,[§] let us suppose a hundred feet off. But as he approached these distant objects, they would dilate before him, though more in the third dimension than superficially, while behind him they would contract. He would know that his eye judged wrongly. If he saw two straight lines which in his estimate ran parallel for the hundred feet to his world's end, he would find on following them that the farther he advanced the more they diverged, because of the dilatation of all the objects to which he approached. On the other hand behind him their distance would seem to diminish, so that as he advanced they would appear always to diverge more and more. But two straight lines which from his first position seemed to converge to one and the same point of the background a hundred feet distant, would continue to do this however far he went, and he would never reach their point of intersection.

Now we can obtain exactly similar images of our real world if we look through a large convex lens of corresponding negative focal length, or even through a pair of convex spectacles if ground somewhat prismatically to resemble pieces of one continuous larger lens. With these, like the convex mirror, we see remote objects as if near to us, the most remote appearing no farther distant than the focus of the lens. In going about with this lens before the eyes, we find that the objects we approach dilate exactly in the manner I have described for pseudospherical space. Now any one using a lens, were it even so strong as to have a focal length of only sixty inches, to say nothing of a hundred feet, would perhaps observe for the first moment that he saw objects brought nearer. But after going about a little the illusion would vanish, and in spite of the false images he would judge of the distances rightly. We have every reason to suppose that what happens

---

[§]The reciprocal of the square of this distance, expressed in negative quantity, would be the measure of curvature of the pseudospherical space.

in a few hours to any one beginning to wear spectacles would soon enough be experienced in pseudospherical space. In short, pseudospherical space would not seem to us very strange, comparatively speaking; we should only at first be subject to illusions in measuring by eye the size and distance of the more remote objects.[14]

There would be illusions of an opposite description, if, with eyes practised to measure in Euclid's space, we entered a spherical space of three dimensions. We should suppose the more distant objects to be more remote and larger than they are, and should find on approaching them that we reached them more quickly than we expected from their appearance. But we should also see before us objects that we can fixate only with diverging lines of sight, namely, all those at a greater distance from us than the quadrant of a great circle. Such an aspect of things would hardly strike us as very extraordinary, for we can have it even as things are if we place before the eye a slightly prismatic glass with the thicker side towards the nose: the eyes must then become divergent to take in distant objects. This excites a certain feeling of unwonted strain in the eyes but does not perceptibly change the appearance of the objects thus seen. The strangest sight, however, in the spherical world would be the back of our own head, in which all visual lines not stopped by other objects would meet again, and which must fill the extreme background of the whole perspective picture.

At the same time it must be noted that as a small elastic flat disc, say of india-rubber, can only be fitted to a slightly curved spherical surface with relative contraction of its border and distension of its centre, so our bodies, developed in Euclid's flat space, could not pass into curved space without undergoing similar distensions and contractions of their parts, their coherence being of course maintained only in as far as their elasticity permitted their bending without breaking. The kind of distension must be the same as in passing from a small body imagined at the centre of Beltrami's sphere to its pseudospherical or spherical representation. For such passage to appear possible, it will always have to be assumed that the body is sufficiently elastic and small in comparison with the real or imaginary radius of curvature of the curved space into which it is to pass.

These remarks will suffice to show the way in which we can infer from the known laws of our sensible perceptions the series of sensible impressions which a spherical or pseudospherical world would give us, if it existed. In doing so we nowhere meet with inconsistency or impossibility any more than in the calculation of its metrical proportions. We can represent to ourselves the look of a pseudospherical world in all directions just as we can develop the conception of it Therefore it cannot be allowed that the axioms of our geometry depend on the native form of our perceptive faculty, or are in ay way connected with it.

It is different with the three dimensions of space. As all our means of sense-perception extend only to space of three dimensions, and a fourth is not merely a modification of what we have but something perfectly new, we find ourselves by reason of our bodily organisation quite unable to represent a fourth dimension.[15]

In conclusion I would again urge that the axioms of geometry are not propositions pertaining only to the pure doctrine of space. As I said before, they are concerned with quantity. We can speak of quantities only when we know of some way by which we can compare, divide and measure them. All space-measurements and therefore in general all ideas of quantities applied to space assume the possibility of figures moving without change of form or size. It is true we are accustomed in geometry to call such figures purely geometrical solids, surfaces, angles and lines, because we abstract from all the other distinctions physical and chemical of natural bodies; but yet one physical quality, rigidity, is retained. Now we have no other mark of rigidity of bodies or figures but congruence, whenever they are applied to one another at any time or place, and after any revolution. We cannot however decide by pure geometry and without mechanical considerations whether the coinciding bodies may not both have varied in the same sense.

If it were useful for any purpose, we might with perfect consistency look upon the space in which we live as the apparent space behind a convex mirror with its shortened and contracted background; or we might consider a bounded sphere of our space, beyond the limits of which we perceive nothing further, as infinite pseudospherical space. Only then we should have to ascribe to the bodies which appear as solid and to our own body at the same time corresponding distensions and contractions, and we must change our system of mechanical principles entirely; for even the proposition that every point in motion, if acted upon by no force, continues to move with unchanged velocity in a straight line, is not adapted to the image of the world in the convex-mirror. The path would indeed be straight, but the velocity would depend upon the place.[16]

Thus the axioms of geometry are not concerned with space relations only but also at the same time with the mechanical deportment of solidest bodies in motion. The notion of rigid geometrical figure might indeed be conceived as transcendental in Kant's sense, namely, as formed independently of actual experience, which need not exactly correspond therewith, any more than natural bodies do ever in fact correspond exactly to the abstract notion we have obtained of them by induction. Taking the notion of rigidity thus as a mere ideal, a strict Kantian might certainly look upon the geometrical axioms as propositions given a priori by transcendental intuition which no experience could either confirm or refute, because it must first be decided by them whether any natural bodies can be considered as rigid. But then we should have to maintain that the axioms of geometry are not synthetic propositions, as Kant held them: they would merely define what qualities and deportment a body must lave to be recognised as rigid.

But if to the geometrical axioms we add propositions relating to the mechanical properties of natural bodies, were it only the axiom of inertia or the single proposition that the mechanical and physical properties of bodies and their mutual reactions are, other circumstances remaining the same, independent of place, such a system of propositions has a real import which can be confirmed or refuted by experience, but just for the same reason can also be got by experience. The mechanical axiom just cited is in fact of the utmost importance for the whole system of our mechanical and physical conceptions. That rigid solids, as we call them, which are really nothing else than elastic solids of great resistance, retain the same form in every part of space if no external force affects them, is a single case falling under the general principle.

For the rest, I do not, of course, suppose that mankind first arrived at space-intuitions in agreement with the axioms of Euclid by any carefully executed systems of exact measurement. It was rather a succession of every day experiences, especially the perception of the geometrical similarity of great and small bodies, only possible in fiat space, that led to the rejection, as impossible, of every geometrical representation at variance with this fact. For this no knowledge of the necessary logical connection between the observed fact of geometrical similarity and the axioms was needed, but only an intuitive apprehension of the typical relations between lines, planes, angles, &c., obtained by numerous and attentive observations—an intuition of the kind the artist possesses of the objects he is to represent, and by means of which he decides surely and accurately whether a new combination which he tries will correspond or not to their nature. It is true that we have no word but *intuition* to mark this; but it is knowledge empirically gained by the aggregation and reinforcement of similar recurrent impressions in memory, and not a transcendental form given before experience. That other such empirical intuitions of fixed typical relations, when not clearly comprehended, have frequently enough been taken by metaphysicians for a priori principles, is a point on which I need not insist.

To sum up, the final outcome of the whole inquiry may be thus expressed:

(1.) The axioms of geometry, taken by themselves out of all connection with mechanical propositions, represent no relations of real things. When thus isolated, if we regard them with Kant as forms of intuition transcendentally given, they constitute a form into which any empirical content whatever will fit and which therefore does not in any way limit or determine beforehand the nature of the content. This is true, however, not only of Euclid's axioms, but also of the axioms of spherical and pseudospherical geometry.

(2.) As soon as certain principles of mechanics are conjoined with the axioms of geometry we obtain a system of propositions which has real import, and which can be verified or overturned by empirical observations, as from experience it can be inferred. If such a system were to be taken as a transcendental form of intuition and thought, there must be assumed a pre-established harmony between form and reality.[17]

Notes

[The text given here is the published English version (Helmholtz 1876), presumably authorized by Helmholtz himself (no translator is listed). Helmholtz notes in a note at the beginning of this essay that "The substance of the first half of the article has been previously expounded by me, in *The Academy* of Feb. 12, 1870 [under the title "The Axioms of Geometry," **1**, 128–131]. It is here set forth anew as necessary context." Helmholtz also published this 1876 version with an additional four paragraphs of introduction (partly addressing Kant's claim that geometrical axioms are synthetic a priori propositions) and a brief appendix about the mathematical description of pseudospherical space in his *Popular Scientific Lectures*, Helmholtz 1962, 223–249, readily available for readers who might wish to compare the later version; there, he notes that this was a lecture delivered in the Docenten Verein in Heidelberg in 1870, which I have taken as the appropriate date for this material. I chose to include this earlier, shorter, version on the grounds that it represented Helmholtz's initial thoughts and also because it has not been republished in this more condensed form.]

1. [Kant is probably meant as the primary "idealistic philosopher" who held this. For Helmholtz's relation to Kant, see Fullinwider 1990.]

2. [To the best of my knowledge, this is the first place in which this fruitful image is introduced, later to become the premise of Edwin Abbott's famous *Flatland* (1992) and many other attempts to imagine curved space or higher dimensions.]

3. [The alternative spelling "geodesic" is currently more common.]

4. [By planimetry Helmholtz means the measurement of geometrical quantities in a Euclidean flat plane.]

5. [By bladder, Helmholtz means in general any empty sac with an aperture, such that the fundus is the side of the bladder opposite to that opening. To understand Helmholtz's example, think of the bladder as roughly hemispherical, usually distended and gathered near the opening, but capable of being folded into a spindle shape (an ordinary balloon may be an even more familiar example).]

6. [For Beltrami's work on the pseudosphere, see Bonola 1955, 127–139.]

7. [Helmholtz here cites Lobachevsky 1829; see also Lobachevsky's "Geometrical Researches on the Theory of Parallels," included at the end of Bonola 1955.]

8. [For references to color theory, see above 34, note 4; the color top was invented in 1849 by J. D. Forbes, Maxwell's teacher; following that, the young Maxwell developed the top to demonstrate how sets of tinted paper when spun could give rise to perceived colors, supporting the three-color theory of color vision. See Everitt 1975, 21 (for a photo of Maxwell and his color top), 65–67.]

9. [Helmholtz was also a great expert on the sensations of tone; specifically, see his essay on "Physiological Causes of Harmony in Music," in Helmholtz 1962, 22–58, at 24; he discusses issues of tone combination in his *magnum opus*, Riemann 1954, 152–233.]

10. [In his 1866 paper given above, Helmholtz calls this property monodromy; see 52 and note 8 in that chapter.]

11. [Implicitly, Helmholtz assumes that the curvature must remain constant, whether positive, negative, or zero.]

12. [Without explicitly stating it, this wonderful thought-experiment seems to give an immediate and intuitive proof that non-Euclidean geometry ought to be as consistent as Euclidean, for every Euclidean proof would have such a "distorted" but faithful non-Euclidean reflection in the curved mirror. Compare this to Poincaré's rather similar train of thought below, 100–101.]

13. [For the details of this mathematical expression, see the Appendix to his later version of this essay, Helmholtz 1962, 247–249, which uses the example of a four-dimensional space whose line element is $ds^2 = dx^2 + dy^2 + dz^2 + dt^2$, though here Helmholtz means by t a fourth spatial variable, not time, as in Einstein's theory.]

14. [An even more commonplace experience, albeit with regard to color perception: when one puts on colored ski goggles, for the first few minutes the whole world seems colored that same hue but very quickly the brain seems to subtract away this extraneous universal hue so that one seems to regain "normal" awareness of colors, or at least no longer notices the goggle hue.]

15. [This categorical assertion should be compared with the different arguments given by Poincaré, below 105.]

16. [This important argument that a force might cause an apparent change in geometry will reemerge in Poincaré, 105–106, below, and Einstein, 1961, 83–86.]

17. [A clear reference to Leibniz's idea that there is a "pre-established harmony" between monads, the "true atoms of nature," each a "simple substance" that "has no windows through which something can enter or leave" so that each must be "a perpetual, living mirror of the universe" and "they agree in virtue of the harmony pre-established between all substances, since they are all representations of a single universe"; Leibniz, "The Monadology," in Leibniz 1989, 213–214, 220-221. Helmholtz is quoting this phrase of Leibniz critically, implying that, were the truths of geometry not based on facts, they would have to be based on a pre-established harmony that defies credulity.]

On the Space-Theory of Matter (1870)

William Kingdon Clifford

Riemann has shown that as there are different kinds of lines and surfaces, so there are different kinds of space of three dimensions; and that we can only find out by experience to which of these kinds the space in which we live belongs. In particular, the axioms of plane geometry are true within the limits of experiment on the surface of a sheet of paper, and yet we know that the sheet is really covered with a number of small ridges and furrows, upon which (the total curvature not being zero) these axioms are not true. Similarly, he says although the axioms of solid geometry are true within the limits of experiment for finite portions of our space, yet we have no reason to conclude that they are true for very small portions; and if any help can be got thereby for the explanation of physical phenomena, we may have reason to conclude that they are not true for very small portions of space.

I wish here to indicate a manner in which these speculations may be applied to the investigation of physical phenomena. I hold in fact

(1) That small portions of space *are* in fact of a nature analogous to little hills on a surface which is on the average flat; namely, that the ordinary laws of geometry are not valid in them.

(2) That this property of being curved or distorted is continually being passed on from one portion of space to another after the manner of a wave.

(3) That this variation of the curvature of space is what really happens in that phenomenon which we call the *motion of matter*, whether ponderable or etherial.

(4) That in the physical world nothing else takes place but this variation, subject (possibly) to the law of continuity.

I am endeavouring in a general way to explain the laws of double refraction on this hypothesis, but have not yet arrived at any results sufficiently decisive to be communicated.

Note

[Clifford 1968, 21–22, there listed as "read Feb. 21, 1870" at the Cambridge Philosophical Society, though it only appeared in their *Proceedings* in 1876, the date often given for this abstract.]

The Postulates of the Science of Space (1873)

William Kingdon Clifford

In my first lecture I said that, out of the pictures which are all that we can really see, we imagine a world of solid things; and that this world is constructed so as to fulfil a certain code of rules, some called axioms, and some called definitions, and some called postulates, and some assumed in the course of demonstration, but all laid down in one form or another in Euclid's *Elements of Geometry.*[1] It is this code of rules that we have to consider to-day. I do not, however, propose to take this book that I have mentioned, and to examine one after another the rules as Euclid has laid them down or unconsciously assumed them; notwithstanding that many things might be said in favour of such a course. This book has been for nearly twenty-two centuries the encouragement and guide of that scientific thought which is one thing with the progress of man from a worse to a better state. The encouragement; for it contained a body of knowledge that was really known and could be relied on, and that moreover was growing in extent and application. For even at the time this book was written—shortly after the foundation of the Alexandrian Museum—Mathematic was no longer the merely ideal science of the Platonic school, but had started on her career of conquest over the whole world of Phenomena. The guide; for the aim of every scientific student of every subject was to bring his knowledge of that subject into a form as perfect as that which geometry had attained. Far up on the great mountain of Truth, which all the sciences hope to scale, the foremost of that sacred sisterhood was seen, beckoning to the rest to follow her. And hence she was called, in the dialect of the Pythagoreans, "the purifier of the reasonable soul." Being thus in itself at once the inspiration and the aspiration of scientific thought, this Book of Euclid's has had a history as chequered as that of human progress itself. It embodied and systematized the truest results of the search after truth that was made by Greek, Egyptian, and Hindu. It presided for nearly eight centuries over that promise of light and right that was made by the civilized Aryan races on the Mediterranean shores; that promise, whose abeyance for nearly as long an interval is so full of warning and of sadness for ourselves. It went into exile along with the intellectual activity and the goodness of Europe. It was taught, and commented upon, and illustrated, and supplemented, by Arab and Nestorian, in the Universities of Baghdad and of Cordova. From these it was brought back into barbaric Europe by terrified students who dared tell hardly any other thing of what they had learned among the Saracens. Translated from Arabic into Latin, it passed into the schools of Europe, spun out with additional cases for every possible variation of the figure, and bristling with words which had sounded to Greek ears like the babbling of birds in a hedge. At length the Greek text appeared and was translated; and, like other Greek authors, Euclid became an authority. There had not yet arisen in Europe "that fruitful faculty," as Mr. Winwood Reade calls it, "with which kindred spirits contemplate each other's works; which not only takes, but gives; which produces from whatever it receives; which embraces to wrestle, and wrestles to embrace." Yet it was coming; and though that criticism of first principles which Aristotle and Ptolemy and Galen underwent waited longer in Euclid's case than in theirs, it came for him at last. What Vesalius was to Galen, what Copernicus was to Ptolemy, that was Lobachevsky to Euclid. There is, indeed, a somewhat instructive parallel between the last two cases. Copernicus and Lobachevsky were both of Slavic origin. Each of them has brought about a revolution in scientific ideas so great that it can only be compared with that wrought by the other. And the reason of the transcendent importance of these two changes is that they are changes in the conception of the Cosmos. Before the time of Copernicus, men knew all about the Universe. They could tell you in the schools, pat off by heart, all that it was, and what it had been, and what it would be. There was the flat earth with the blue vault of heaven resting on it like the dome of a cathedral, and the bright cold stars stuck into it; while the sun and planets moved in crystal spheres between. Or, among the better informed, the earth was a globe in the centre of the universe, heaven a sphere concentric with it; intermediate machinery as before. At any rate, if there was anything beyond heaven, it was a void space that needed no further description. The history of all this could be traced back to a certain definite time, when it began; behind that was a changeless eternity that needed no further history. Its future could be predicted in general terms as far forward as a certain epoch, about the precise determination of which there were, indeed, differences among the learned. But after that would come again a changeless eternity, which was fully accounted for and described. But in any case the Universe was a known thing. Now the enormous effect of the Copernican system, and of the astronomical discoveries that have followed it, is that, in place of this knowledge of a little, which was called knowledge of the Universe, of Eternity and Immensity, we have now got knowledge of a great deal more; but we only call it the knowledge of Here and Now. We can tell a great deal about the solar system; but, after all, it is our house, and not the city. We can tell something about the star-system to which our sun belongs; but, after all, it is our star-system, and not the Universe. We are talking about Here with the consciousness of a There beyond it, which we may know some time, but do not at all know now. And though the nebular hypothesis tells us a great deal about the history of the solar system, and traces it back for a period compared with which the old measure of the duration of the Universe from beginning to end is not a second to a century, yet we do not call this the history of eternity.[2] We may put it all together and call it Now, with the consciousness of a Then before it, in which things were happening that may have left records; but we have not yet read them. This, then, was the change effected by Copernicus in the idea of the Universe. But there was left another to be made. For the laws of space and motion, that we are presently going to examine, implied an infinite space and an infinite duration, about whose properties as space and time everything was accurately known. The very constitution of those parts of it which are at an infinite distance from us, "geometry upon the plane at infinity," is just as well known, if the Euclidean assumptions are true, as the geometry of any portion of this room. In this infinite and thoroughly well-known space the Universe is situated during at least some portion of an infinite and thoroughly well-known time. So that here we have real knowledge of something at least that concerns the Cosmos; something that is true throughout the Immensities and the Eternities. That something Lobachevsky and his successors have taken away. The geometer of to-day knows nothing about the nature of actually existing space at an infinite distance; he knows nothing about the properties of this present space in a past or a future eternity. He knows, indeed, that the laws assumed by Euclid are true with an accuracy that no direct experiment can approach, not only in this place where we are, but in places at a distance from us that no astronomer has conceived; but he knows this as of Here and Now; beyond his range is a There and Then of which he knows nothing at present, but may ultimately come to know more. So, you see, there is a real parallel between the work of Copernicus and his successors on the one hand, and the work of Lobachevsky and his successors on the other. In both of these the knowledge of Immensity and Eternity is replaced by knowledge of Here and Now. And in virtue of these two revolutions the idea of the Universe, the Macrocosm, the All, as subject of human knowledge, and therefore of human interest, has fallen to pieces.

It will now, I think, be clear to you why it will not do to take for our present consideration the postulates of geometry as Euclid has laid them down. While they were all certainly true, there might be substituted for them some other group of equivalent propositions; and the choice of the particular set of statements that should be used as the groundwork of the science was to a certain extent arbitrary, being only guided by convenience of exposition. But from the moment that the actual truth of these assumptions becomes doubtful, they fall of themselves into a necessary order and classification; for we then begin to see which of them may be true independently of the others. And for the purpose of criticizing the evidence for them, it is essential that this natural order should be taken; for I think you will see presently that any other order would bring hopeless confusion into the discussion.

Space is divided into parts in many ways. If we consider any material thing, space is at once divided into the part where that thing is and the part where it is not.[3] The water in this glass, for example, makes a distinction between the space where it is and the space where it is not. Now, in order to get from one of these to the other you must cross the *surface* of the water; this surface is the boundary of the space where the water is which separates it from the space where it is not. Every *thing*, considered as occupying a portion of space, has a surface which separates the space where it is from the space where it is not. But, again, a surface may be divided into parts in various ways. Part of the surface of this water is against the air, and part is against the glass. If you travel over the surface from one of these parts to the other, you have to cross the *line* which divides them; it is this circular edge where water, air, and glass meet. Every part of a surface is separated from the other parts by a line which bounds it. But now suppose, further, that this glass had been so constructed that the part towards you was blue and the part towards me was white, as it is now. Then this line, dividing two parts of the surface of the water, would itself be divided into two parts; there would be a part where it was against the blue glass, and a part where it was against the white glass. If you travel in thought along that line, so as to get from one of these two parts to the other, you have to cross a *point* which separates them, and is the boundary between them. Every part of a line is separated from the other parts by points which bound it. So we may say altogether—The boundary of a solid (i.e., of a part of space) is a surface.

The boundary of a part of a surface is a line.

The boundaries of a part of a line are points.

And we are only settling the meanings in which words are to be used. But here we may make an observation which is true of all space that we are acquainted with: it is that the process ends here. There are no parts of a point which are separated from one another by the next link in the series. This is also indicated by the reverse process.

For I shall now suppose this point—the last thing that we got to—to move round the tumbler so as to trace out the line, or edge, where air, water, and glass meet. In this way I get a series of points, one after another; a series of such a nature that, starting from any one of them, only two changes are possible that will keep it within the series: it must go forwards or it must go backwards, and each of these is perfectly definite. The line may then be regarded as an aggregate of points. Now let us imagine, further, a change to take place in this line, which is nearly a circle. Let us suppose it to contract towards the centre of the circle, until it becomes indefinitely small, and disappears. In so doing it will trace out the upper surface of the water, the part of the surface where it is in contact with the air. In this way we shall get a series of circles one after another—a series of such a nature that, starting from any one of them, only two changes are possible that will keep it within the series: it must expand or it must contract. This series, therefore, of circles, is just similar to the series of points that make one circle; and just as the line is regarded as an aggregate of points, so we may regard this surface as an aggregate of lines. But this surface is also in another sense an aggregate of points, in being an aggregate of aggregates of points. But, starting from a point in the surface, more than two changes are possible that will keep it within the surface, for it may move in any direction. The surface, then, is an aggregate of points of a different kind from the line. We speak of the line as a point-aggregate of one dimension, because, starting from one point, there are only two possible directions of change; so that the line can be traced out in one motion. In the same way, a surface is a line-aggregate of one dimension, because it can be traced out by one motion of the line; but it is a point-aggregate of two dimensions, because, in order to build it up of points, we have first to aggregate points into a line, and then lines into a surface. It requires two motions of a point to trace it out.

Lastly, let us suppose this upper surface of the water to move downwards, remaining always horizontal till it becomes the under surface. In so doing it will trace out the part of space occupied by the water. We shall thus get a series of surfaces one after another, precisely analogous to the series of points which make a line, and the series of lines which make a surface. The piece of solid space is an aggregate of surfaces, and an aggregate of the same kind as the line is of points; it is a surface-aggregate of one dimension. But at the same time it is a line-aggregate of two dimensions, and a point-aggregate of three dimensions. For if you consider a particular line which has gone to make this solid, a circle partly contracted and part of the way down, there are more than two opposite changes which it can undergo. For it can ascend or descend, or expand or contract, or do both together in any proportion. It has just as great a variety of changes as a point in a surface. And the piece of space is called a point-aggregate of three dimensions, because it takes three distinct motions to get it from a point. We must first aggregate points into a line, then lines into a surface, then surfaces into a solid.

At this step it is clear, again, that the process must stop in all the space we know of. For it is not possible to move that piece of space in such a way as to change every point in it. When we moved our line or our surface, the new line or surface contained no point whatever that was in the old one; we started with one aggregate of points, and by moving it we got an entirely new aggregate, all the points of which were new. But this cannot be done with the solid; so that the process is at an end. We arrive, then, at the result that *space is of three dimensions.*

Is this, then, one of the postulates of the science of space? No; it is not. The science of space, as we have it, deals with relations of distance existing in a certain space of three dimensions, but it does not at all require us to assume that no relations of distance are possible in aggregates of more than three dimensions. The fact that there are only three dimensions does regulate the number of books that we write, and the parts of the subject that we study: but it is not itself a postulate of the science. We investigate a certain space of three dimensions, on the hypothesis that it has certain elementary properties; and it is the assumptions of these elementary properties that are the real postulates of the science of space. To these I now proceed.

The first of them is concerned with *points*, and with the relation of space to them. We spoke of a line as an aggregate of points. Now there are two kinds of aggregates, which are called respectively continuous and discrete. If you consider this line, the boundary of part of the surface of the water, you will find yourself believing that between any two points of it you can put more points of division, and between any two of these more again, and so on; and you do not believe there can be any end to the process. We may express that by saying you believe that between any two points of the line there is an infinite number of other points. But now here is an aggregate of marbles, which, regarded as an aggregate, has many characters of resemblance with the aggregate of points. It is a series of marbles, one after another; and if we take into account the relations of nextness or contiguity which they possess, then there are only two changes possible from one of them as we travel along the series: we must go to the next in front, or to the next behind. But yet it is not true that between any two of them here is an infinite number of other marbles; between these two, for example, there are only three. There, then, is a distinction at once between the two kinds of aggregates. But there is another, which was pointed out by Aristotle in his Physics and made the basis of a definition of continuity. I have here a row of two different kinds of marbles, some white and some black. This aggregate is divided into two parts, as we formerly supposed the line to be. In the case of the line the boundary between the two parts is a point which is the element of which the line is an aggregate. In this case before us, a marble is the element; but here we cannot say that the boundary between the two parts is a marble. The boundary of the white parts is a white marble, and the boundary of the black parts is a black marble; these two adjacent parts have different boundaries. Similarly, if instead of arranging my marbles in a series, I spread them out on a surface, I may have this aggregate divided into two portions—a white portion and a black portion; but the boundary of the white portion is a row of white marbles, and the boundary of the black portion is a row of black marbles. And lastly, if I made a heap of white marbles, and put black marbles on the top of them, I should have a discrete aggregate of three dimensions divided into two parts: the boundary of the white part would be a layer of white marbles, and the boundary of the black part would be a layer of black marbles. In all these cases of discrete aggregates, when they are divided into two parts, the two adjacent parts have different boundaries. But if you come to consider an aggregate that you believe to be continuous, you will see that you think of two adjacent parts as having the *same* boundary. What is the boundary between water and air here? Is it water? No; for there would still have to be a boundary to divide that water from the air. For the same reason it cannot be air. I do not want you at present to think of the actual physical facts by the aid of any molecular theories; I want you only to think of what appears to be, in order to understand clearly a conception that we all have. Suppose the things actually in contact. If, however much we magnified them, they still appeared to be thoroughly homogeneous, the water filling up a certain space, the air an adjacent space; if this held good indefinitely through all degrees of conceivable magnifying, then we could not say that the surface of the water was a layer of water and the surface of air a layer of air; we should have to say that the same surface was the surface of both of them, and was itself neither one nor the other—that this surface occupied no space at all. Accordingly, Aristotle defined the continuous as that of which two adjacent parts have the same boundary; and the discontinuous or discrete as that of which two adjacent parts have direct boundaries.*

---

* *Physics* V.iii [227a10–14] "A thing that is in succession and touches is contiguous. The continuous is a subdivision of the contiguous: things are called continuous when the touching limits of each become one and the same and are, as the word implies, contained in each

Now the first postulate of the science of space is that space is a continuous aggregate of points, and not a discrete aggregate. And this postulate—which I shall call the postulate of continuity—is really involved in those three of the six[†] postulates of Euclid for which Robert Simson has retained the name of postulate.[4] You will see, on a little reflection, that a discrete aggregate of points could not be so arranged that any two of them should be relatively situated to one another in exactly the same manner, so that any two points might be joined by a straight line which should always bear the same definite relation to them. And the same difficulty occurs in regard to the other two postulates. But perhaps the most conclusive way of showing that this postulate is really assumed by Euclid is to adduce the proposition he proves, that every finite straight line may be bisected. Now this could not be the case if it consisted of an odd number of separate points. As the first of the postulates of the science of space, then, we must reckon this postulate of Continuity; according to which two adjacent portions of space, or of a surface, or of a line, have the *same* boundary, viz.—a surface, a line, or a point; and between every two points on a line there is an infinite number of intermediate points.

The next postulate is that of Elementary Flatness. You know that if you get hold of a small piece of a very large circle, it seems to you nearly straight. So, if you were to take any curved line, and magnify it very much, confining your attention to a small piece of it, that piece would seem straighter to you than the curve did before it was magnified. At least, you can easily conceive a curve possessing this property, that the more you magnify it, the straighter it gets. Such a curve would possess the property of elementary flatness. In the same way, if you perceive a portion of the surface of a very large sphere, such as the earth, it appears to you to be flat. If, then, you take a sphere of say a foot diameter, and magnify it more and more, you will find that the more you magnify it the flatter it gets. And you may easily suppose that this process would go on indefinitely; that the curvature would become less and less the more the surface was magnified. Any curved surface which is such that the more you magnify it the flatter it gets, is said to possess the property of elementary flatness. But if every succeeding power of our imaginary microscope disclosed new wrinkles and inequalities without end, then we should say that the surface did not possess the property of elementary flatness.

But how am I to explain how solid space can have this property of elementary flatness? Shall I leave it as a mere analogy, and say that it is the same kind of

---

other: continuity is impossible if these extremities are two." A little further on he makes the important remark that on the hypothesis of continuity a line is not *made up* of points in the same way that a whole is made up of parts, VI.i [231a24–26] "nothing that is continuous can be composed of indivisibles: e.g. a line cannot be composed of points, the line being continuous and the point indivisible." [Clifford cites these passages in Greek; the translation used here is from Aristotle 1984, 1:384, 390–391.]

[†]See De Morgan, in Smith's Dict. of Biography and Mythology, Art. *Euclid*; and in the English Cyclopdia, Art. *Axiom*

property as this of the curve and surface, only in three dimensions instead of one or two? I think I can get a little nearer to it than that; at all events I will try.

If we start to go out from a point on a surface, there is a certain choice of directions in which we may go. These directions make certain angles with one another. We may suppose a certain direction to start with, and then gradually alter that by turning it round the point: we find thus a single series of directions in which we may start from the point. According to our first postulate, it is a continuous series of directions. Now when I speak of a direction from the point, I mean a direction of starting; I say nothing about the subsequent path. Two different paths may have the same direction at starting; in this case they will touch at the point; and there is an obvious difference between two paths which touch and two paths which meet and form an angle. Here, then, is an aggregate of directions, and they can be changed into one another. Moreover, the changes by which they pass into one another have magnitude, they constitute distance-relations; and the amount of change necessary to turn one of them into another is called the angle between them. It is involved in this postulate that we are considering, that angles can be compared in respect of magnitude. But this is not all. If we go on changing a direction of start, it will, after a certain amount of turning, come round into itself again, and be the same direction. On every surface which has the property of elementary flatness, the amount of turning necessary to take a direction all round into its first position is the same for all points of the surface. I will now show you a surface which at one point of it has not this property. I take this circle of paper from which a sector has been cut out, and bend it round so as to join the edges; in this way I form a surface which is called a *cone*. Now on all points of this surface but one, the law of elementary flatness holds good. At the vertex of the cone, however, notwithstanding that there is an aggregate of directions in which you may start, such that by continuously changing one of them you may get it round into its original position, yet the whole amount of change necessary to effect this is not the same at the vertex as it is at any other point of the surface. And this you can see at once when I unroll it; for only part of the directions in the plane have been included in the cone. At this point of the cone, then, it does not possess the property of elementary flatness; and no amount of magnifying would ever make a cone seem flat at its vertex.

To apply this to solid space, we must notice that here also there is a choice of directions in which you may go out from any point; but it is a much greater choice than a surface gives you. Whereas in a surface the aggregate of directions is only of one dimension, in solid space it is of two dimensions. But here also there are distance-relations, and the aggregate of directions may be divided into parts which have quantity. For example, the directions which start from the vertex of this cone are divided into those which go inside the cone, and those which go outside the cone. The part of the aggregate which is inside the cone is called a solid angle. Now in those spaces of three dimensions which have the property of elementary flatness, the whole amount of solid angle round one point is equal to the whole amount round another point. Although the space need not be exactly similar to itself in all parts, yet the aggregate of directions round one point is exactly similar to the aggregate of directions round another point, if the space has the property of elementary flatness.

How does Euclid assume this postulate of Elementary Flatness? In his fourth postulate he has expressed it so simply and clearly that you will wonder how anybody could make all this fuss. He says, "All right angles are equal."

Why could I not have adopted this at once, and saved a great deal of trouble? Because it assumes the knowledge of a surface possessing the property of elementary flatness in all its points. Unless such a surface is first made out to exist, and the definition of a right angle is restricted to lines drawn upon it—for there is no necessity for the word *straight* in that definition—the postulate in Euclid's form is obviously not true. I can make two lines cross at the vertex of a cone so that the four adjacent angles shall be equal, and yet not one of them equal to a right angle.

I pass on to the third postulate of the science of space—the postulate of Superposition. According to this postulate a body can be moved about in space without altering its size or shape.[5] This seems obvious enough, but it is worth while to examine a little closely into the meaning of it. We must define what we mean by size and by shape. When we say that a body can be moved about without altering its size, we mean that it can be so moved as to keep unaltered the length of all the lines in it. This postulate therefore involves that lines can be compared in respect of magnitude, or that they have a length independent of position; precisely as the former one involved the comparison of angular magnitudes. And when we say that a body can be moved about without altering its shape, we mean that it can be so moved as to keep unaltered all the angles in it. It is not necessary to make mention of the motion of a body, although that is the easiest way of expressing and of conceiving this postulate; but we may, if we like, express it entirely in terms which belong to space, and that we should do in this way. Suppose a figure to have been constructed in some portion of space; say that a triangle has been drawn whose sides are the shortest distances between its angular points. Then if in any other portion of space two points are taken whose shortest distance is equal to a side of the triangle, and at one of them an angle is made equal to one of the angles adjacent to that side, and a line of shortest distance drawn equal to the corresponding side of the original triangle, the distance from the extremity of this to the other of the two points will be equal to the third side of the original triangle, and the two will be equal in all respects; or generally, if a figure has been constructed anywhere, another figure, with all its lines and all its angles equal to the corresponding lines and angles of the first, can be constructed anywhere else. Now this is exactly what is meant by the principle of superposition employed by Euclid to prove the proposition that I have just mentioned. And we may state it again in this short form—All parts of space are exactly alike.

But this postulate carries with it a most important consequence. It enables us to make a pair of most fundamental definitions—those of the plane and of the straight line. In order to explain how these come out of it when it is granted, and how they cannot be made when it is not granted, I must here say something more about the nature of the postulate itself, which might otherwise have been left until we come to criticize it.

We have stated the postulate as referring to solid space. But a similar property may exist in surfaces. Here, for instance, is part of the surface of a sphere. If I draw any figure I like upon this, I can suppose it to be moved about in any way upon the sphere, without alteration of its size or shape. If a figure has been drawn on any part of the surface of a sphere, a figure equal to it in all respects may be drawn on any other part of the surface. Now I say that this property belongs to the surface itself, is a part of its own internal economy, and does not depend in any way upon its relation to space of three dimensions. For I can pull it about and bend it in all manner of ways, so as altogether to alter its relation to solid space; and yet, if I do not stretch it or tear it, I make no difference whatever in the length of any lines upon it, or in the size of any angles upon it.[‡] I do not in any way alter the figures drawn upon it, or the possibility of drawing figures upon it, *so far as their relations with the surface itself are concerned.* This property of the surface, then, could be ascertained by people who lived entirely in it, and were absolutely ignorant of a third dimension. As a point-aggregate of two dimensions, it has in itself properties determining the distance-relations of the points upon it, which are absolutely independent of the existence of any points which are not upon it.

Now here is a surface which has not that property. You observe that it is not of the same shape all over, and that some parts of it are more curved than other parts. If you drew a figure upon this surface, and then tried to move it about, you would find that it was impossible to do so without altering the size and shape of the figure. Some parts of it would have to expand, some to contract, the lengths of the lines could not all be kept the same, the angles would not hit off together. And this property of the surface—that its parts are different from one another—is a property of the surface itself, a part of its internal economy, absolutely independent of any relations it may have with space outside of it. For, as with the other one, I can pull it about in all sorts of ways, and, so long as I do not stretch it or tear it, I make no alteration in the length of lines drawn upon it or in the size of the angles.

Here, then, is an intrinsic difference between these two surfaces, as surfaces. They are both point-aggregates of two dimensions; but the points in them have certain relations of distance (distance measured always *on* the surface), and these relations of distance are not the same in one case as they are in the other.

The supposed people living in the surface and having no idea of a third dimension might, without suspecting that third dimension at all, make a very

[‡]This figure was made of linen, starched upon a spherical surface, and taken off when dry. That mentioned in the next paragraph was similarly stretched upon the irregular surface of the head of a bust. For durability these models should be made of two thicknesses of linen starched together in such a way that the fibres of one bisect the angles between the fibres of the other, and the edge should be bound by a thin slip of paper. They will then retain their curvature unaltered for a long time.

accurate determination of the nature of their *locus in quo*. If the people who lived on the surface of the sphere were to measure the angles of a triangle, they would find them to exceed two right angles by a quantity proportional to the area of the triangle. This excess of the angles above two right angles, being divided by the area of the triangle, would be found to give exactly the same quotient at all parts of the sphere. That quotient is called the curvature of the surface; and we say that a sphere is a surface of uniform curvature. But if the people living on this irregular surface were to do the same thing, they would not find quite the same result. The sum of the angles would, indeed, differ from two right angles, but sometimes in excess, and sometimes in defect, according to the part of the surface where they were. And though for small triangles in any one neighbourhood the excess or defect would be nearly proportional to the area of the triangle, yet the quotient obtained by dividing this excess or defect by the area of the triangle would vary from one part of the surface to another. In other words, the curvature of this surface varies from point to point; it is sometimes positive, sometimes negative, sometimes nothing at all.

But now comes the important difference. When I speak of a triangle, what do I suppose the sides of that triangle to be?

If I take two points near enough together upon a surface, and stretch a string between them, that string will take up a certain definite position upon the surface, marking the line of shortest distance from one point to the other. Such a line is called a geodesic line. It is a line determined by the intrinsic properties of the surface, and not by its relations with external space. The line would still be the shortest line, however the surface were pulled about without stretching or tearing. A geodesic line may be *produced*, when a piece of it is given; for we may take one of the points, and, keeping the string stretched, make it go round in a sort of circle until the other end has turned through two right angles. The new position will then be a prolongation of the same geodesic line.

In speaking of a triangle, then, I meant a triangle whose sides are geodesic lines. But in the case of a spherical surface—or, more generally, of a surface of constant curvature—these geodesic lines have another and most important property. They are *straight*, so far as the surface is concerned. On this surface a figure may be moved about without altering its size or shape. It is possible, therefore, to draw a line which shall be of the same shape all along and on both sides. That is to say, if you take a piece of the surface on one side of such a line, you may slide it all along the line and it will fit; and you may turn it round and apply it to the other side, and it will fit there also. This is Leibniz's definition of a straight line, and, you see, it has no meaning except in the case of a surface of constant curvature, a surface all parts of which are alike.

Now let us consider the corresponding things in solid space. In this also we may have geodesic lines; namely, lines formed by stretching a string between two points. But we may also have geodesic surfaces; and they are produced in this manner. Suppose we have a point on a surface, and this surface possesses the property of elementary flatness. Then among all the directions of starting from the point, there are some which start *in the surface*, and do not make an angle with it. Let all these be prolonged into geodesics; then we may imagine one of these geodesics to travel round and coincide with all the others in turn. In so doing it will trace out a surface which is called a geodesic surface. Now in the particular case where a space of three dimensions has the property of superposition, or is all over alike, these geodesic surfaces are *planes*. That is to say, since the space is all over alike, these surfaces are also of the same shape all over and on both sides; which is Leibniz's definition of a plane. If you take a piece of space on one side of such a plane, partly bounded by the plane, you may slide it all over the plane, and it will fit; and you may turn it round and apply it to the other side, and it will fit there also. Now it is clear that this definition will have no meaning unless the third postulate be granted. So we may say that when the postulate of Superposition is true, then there are planes and straight lines; and they are defined as being of the same shape throughout and on both sides.

It is found that the whole geometry of a space of three dimensions is known when we know the curvature of three geodesic surfaces at every point. The third postulate requires that the curvature of all geodesic surfaces should be everywhere equal to the same quantity.

I pass to the fourth postulate, which I call the postulate of Similarity. According to this postulate, any figure may be magnified or diminished in any degree without altering its shape. If any figure has been constructed in one part of space, it may be reconstructed to any scale whatever in any other part of space, so that no one of the angles shall be altered though all the lengths of lines will of course be altered. This seems to be a sufficiently obvious induction from experience; for we have all frequently seen different sizes of the same shape; and it has the advantage of embodying the fifth and sixth of Euclid's postulates in a single principle, which bears a great resemblance in form to that of Superposition, and may be used in the same manner. It is easy to show that it involves the two postulates of Euclid: "Two straight lines cannot enclose a space," and "Lines in one plane which never meet make equal angles with every other line."

This fourth postulate is equivalent to the assumption that the constant curvature of the geodesic surfaces is zero; or the third and fourth may be put together, and we shall then say that the three curvatures of space are all of them zero at every point.

The supposition made by Lobachevsky was, that the three first postulates were true, but not the fourth. Of the two Euclidean postulates included in this, he admitted one, viz., that two straight lines cannot enclose a space, or that two lines which once diverge go on diverging for ever. But he left out the postulate about parallels, which may be stated in this form. If through a point outside of a straight line there be drawn another, indefinitely produced both ways; and if we turn this second one round so as to make the point of intersection travel along the first line, then at the very instant that this point of intersection disappears at one end it will reappear at the other, and there is only one position in which the lines do not intersect. Lobachevsky supposed, instead, that there was a finite angle through which the second line must be turned after the point of intersection had disappeared at one end, before it reappeared at the other. For all positions of the second line within this angle there is then no intersection. In the two limiting positions, when the lines have just done meeting at one end, and when they are just going to meet at the other, they are called parallel; so that two lines can be drawn through a fixed point parallel to a given straight line. The angle between these two depends in a certain way upon the distance of the point from the line. The sum of the angles of a triangle is less than two right angles by a quantity proportional to the area of the triangle. The whole of this geometry is worked out in the style of Euclid, and the most interesting conclusions are arrived at; particularly in the theory of solid space, in which a surface turns up which is not plane relatively to that space, but which, for purposes of drawing figures upon it, is identical with the Euclidean plane.

It was Riemann, however, who first accomplished the task of analysing all the assumptions of geometry, and showing which of them were independent. This very disentangling and separation of them is sufficient to deprive them for the geometer of their exactness and necessity; for the process by which it is effected consists in showing the possibility of conceiving these suppositions one by one to be untrue; whereby it is clearly made out how much is supposed. But it may be worth while to state formally the case for and against them.

When it is maintained that we know these postulates to be universally true, in virtue of certain deliverances of our consciousness, it is implied that these deliverances could not exist, except upon the supposition that the postulates are true. If it can be shown, then, from experience that our consciousness would tell us exactly the same things if the postulates are not true, the ground of their validity will be taken away. But this is a very easy thing to show.

That same faculty which tells you that space is continuous tells you that this water is continuous, and that the motion perceived in a wheel of life[6] is continuous. Now we happen to know that if we could magnify this water as much again as the best microscopes can magnify it, we should perceive its granular structure. And what happens in a wheel of life is discovered by stopping the machine. Even apart, then, from our knowledge of the way nerves act in carrying messages, it appears that we have no means of knowing anything more about an aggregate than that it is too fine-grained for us to perceive its discontinuity, if it has any.

Nor can we, in general, receive a conception as positive knowledge which is itself founded merely upon inaction. For the conception of a continuous thing is of that which looks just the same however much you magnify it. We may conceive the magnifying to go on to a certain extent without change, and then, as it were, leave it going on, without taking the trouble to doubt about the changes that may ensue.

In regard to the second postulate, we have merely to point to the example of polished surfaces. The smoothest surface that can be made is the one most completely covered with the minutest ruts and furrows. Yet geometrical constructions can be made with extreme accuracy upon such a surface, on the supposition that it is an exact plane. If, therefore, the sharp points, edges, and furrows of space are only small enough, there will be nothing to hinder our conviction of its elementary flatness. It has even been remarked by Riemann that we must not shrink from this supposition if it is found useful in explaining physical phenomena.

The first two postulates may therefore be doubted on the side of the very small. We may put the third and fourth together, and doubt them on the side of the very great. For if the property of elementary flatness exist on the average, the deviations from it being, as we have supposed, too small to be perceived, then, whatever were the true nature of space, we should have exactly the conceptions of it which we now have, if only the regions we can get at were small in comparison with the areas of curvature. If we suppose the curvature to vary in an irregular manner, the effect of it might be very considerable in a triangle formed by the nearest fixed stars; but if we suppose it approximately uniform to the limit of telescopic reach, it will be restricted to very much narrower limits. I cannot perhaps do better than conclude by describing to you as well as I can what is the nature of things on the supposition that the curvature of all space is nearly uniform and positive.

In this case the Universe, as known, becomes again a valid conception; for the extent of space is a finite number of cubic miles.[§] And this comes about in a curious way. If you were to start in any direction whatever, and move in that direction in a perfect straight line according to the definition of Leibniz; after travelling a most prodigious distance, to which the parallactic unit[7]—200,000 times the diameter of the earth's orbit—would be only a few steps, you would arrive at—this place. Only, if you had started upwards, you would appear from below. Now, one of two things would be true. Either, when you had got half-way on your journey, you came to a place that is opposite to this, and which you must have gone through, whatever direction you started in; or else all paths you could have taken diverge entirely from each other till they meet again at this place. In the former case, every two straight lines in a plane meet in two points, in the latter they meet only in one. Upon this supposition of a positive curvature, the whole of geometry is far more complete and interesting; the principle of duality, instead of half breaking down over metric relations, applies to all propositions without exception. In fact, I do not mind confessing that I personally have often found relief from the dreary infinities of homaloidal[8] space in the consoling hope that, after all, this other may be the true state of things.

[§]The assumptions here made about the *Zusammenhang* [continuity] of space are the simplest ones, but even the finite extent does not follow necessarily from uniform positive curvature; as Riemann seems to have supposed.

Notes

[This was the third in a series of lectures Clifford delivered in 1873 at the Royal Institution in London under the general title "The Philosophy of the Pure Sciences"; the text is taken from Clifford 1886, 210–230, which I have left as it stands except for regularizing the spelling of names such as Lobachevsky and Leibniz.]

1. [Clifford's previous lectures were explorations of the reality of the external word, leading to his defense of Kant's view that "every apparently universal statement ... is a particular statement about my nervous system, about my apparatus of thought; or that I do not know that it is true" (Clifford 1886, 200).]

2. [The Kant-Laplace nebula hypothesis (announced by Kant in 1755) holds that solar systems such as our own are formed from contracting and rotating clouds of gas; see Harrison 1987, 108. This hypothesis remains the basis of our present understanding of these processes.]

3. [Clifford gives a more extended discussion of these issues in his essay "Of Boundaries in General," in Clifford 1879, 127–156; this book also presents Clifford's account of the role of the brain and eye in the larger context of seeing and thinking.]

4. [Robert Simson (1687–1768) prepared an edition of Euclid's *Elements* (1756) that went through many editions and greatly influenced most modern English versions of Euclid; he defended Euclid's axioms as not needing revision or addition. Simson also published "restorations" of lost works by Euclid and Apollonius; see Boyer and Merzbach 1991, 459.]

5. [Clifford's postulate of superposition is essentially Helmholtz's requirement of the free mobility and rotation of rigid bodies. The requirement that angles remain unchanged is now called *conformal invariance*.]

6. [In 1834, William Horner invented what he called named the "Daedalum" ("wheel of the Devil"), later reinvented by others and called the "wheel of life" or "zoetrope," in which a series of hand-drawn images attached to the inside of a cylindrical drum appeared to be animated when the drum was rotated. Slits in the sides of the drum allow several observers to see the animation simultaneously. Such optical toys were the precursors of the cinema and relied on the same basic principle of the persistence of images.]

7. [By "parallactic unit," Clifford means what is now called a *parsec*, 3.1×10^{16} m, the distance at which a star is observed to have a parallax of one second of arc, meaning the angle subtended when the star is viewed from two points separated by one astronomical unit (the mean distance between the earth and the sun).]

8. [By "homaloidal," Clifford means Euclidean.]

Elementary Theorems Relating to the Geometry of a Space of Three Dimensions and of Uniform Positive Curvature in the Fourth Dimension (1877)

Simon Newcomb

The following theorems are founded on the ideas of *Riemann*, as set forth in his celebrated dissertation "Ueber die Hypothesen, welche der Geometrie zu Grunde liegen," though they may not be entirely accordant with his remarks on the result of his theory. It appears not uninteresting to consider the subject from the standpoint of elementary geometry instead of following the analytic method which has been commonly employed by writers on the non-Euclidean geometry. The system here set forth is founded on the following three postulates.

1. I assume that space is triply extended, unbounded, without properties dependent either upon position or direction, and possessing such planeness in its smallest parts that both the postulates of the Euclidean geometry, and our common conceptions of the relations of the parts of space are true for every indefinitely small region in space.

2. I assume that this space is affected with such curvature that a right line[1] shall always return into itself at the end of a finite and real distance $2D$ without losing, in any part of its course, that symmetry with respect to space on all sides of it which constitutes the fundamental property of our conception of it.

3. I assume that if two right lines emanate from the same point, making the indefinitely small angle α with each other, their distance apart at the distance r from the point of intersection will be given by the equation[2]

$$s = \frac{2\alpha D}{\pi} \sin \frac{r\pi}{2D}. \qquad [1]$$

The right line thus has this property in common with the Euclidean right line that two such lines intersect in only a single point. It may be that the number of points in which two such lines can intersect admits of being determined from the laws of curvature, but not being able so to determine it, I assume as a postulate the fundamental property of the Euclidean right line. It remains to be seen whether this assumption leads to any conclusions either inconsistent with themselves or to the Euclidean geometry in any small region of space. The following nomenclature may be used. A *complete right line* is one returning into itself as supposed in postulate 2. Any small portion of it is to be conceived of as a Euclidean right line. It may be called a right line simply when no ambiguity will result therefrom.

The locus of all complete right lines passing through the same point, and lying in the same Euclidean plane containing that point will be called a *complete plane.*

A *region* will mean any indefinitely small portion of space, in which we are to conceive of the Euclidean geometry as holding true. Within any region whatever figures may be designated as Euclidean in order to avoid confusing them with the more complicated relations which have place in the geometry of curved space.

The following propositions are for the most part, presented without demonstration as being either too obvious to require it or obtainable by processes which leave no doubt of their validity. A few will need at least the outlines of a demonstration.

I. From postulates 1 and 2 it follows that *all complete right lines are of the same length* $2D$. Hence D is the greatest possible distance at which any two points in space can be situated, it being supposed that the distance is measured on the line of least absolute length. If two moving points start out in opposite directions from a point A on a right line α, they will meet at the distance D in a point which we may designate as $A'\alpha$.

II. *The complete plane is a Euclidean plane in every region of its extent.* For, let α, α', and α'' be three successive positions of the generating right line, and let r, r', and r'' be three points each at any distance r from the common point of intersection of the lines α, α', and α''. Then, considering the Euclidean plane containing the line α and the point r', there can, owing to the symmetry of space on each side (postulate 1) be no reason why the line α' should intersect this plane in one direction rather than in another, it will therefore wholly lie in it. And, from the same postulate, there is no reason why the line α'' should pass on one side of the plane $\alpha r'$ rather than on the other; it will therefore lie in it. Therefore, in every region, the consecutive positions of the generating line lie in the same Euclidean plane.

III. *Every system of right lines, passing through a common point A and making an indefinitely small angle with each other, are parallel to each other in the region A' at distance D.* From postulate 3 it follows that in this region we have $\frac{ds}{dr} = 0$,[3] while, by proposition II, every pair lie in the same plane. Conversely, since two points completely determine a right line, it follows that *all lines which are parallel in the same region intersect in a common point at the distance D from that region.*

IV. *If a system of right lines pass in the same plane through A, the locus of their most distant points will be a complete right line.* It is obvious that this locus will be everywhere perpendicular to the generating line, because there is no reason why the angle on one side should be different from that on the other. Moreover, there is no reason why the locus at any point should deviate to one side of the Euclidean plane containing two consecutive positions of the generating line rather than to the other. It will therefore, in every region, be a

Euclidean right line. And, when the generating line has turned through 180°, the most distant point will have returned to its original position: it will therefore have described a complete right line.

V. *The locus of all the points at distance D from a fixed point A, is a complete plane, and, indeed, a double plane if we consider as distinct the coincident surfaces in which the two opposite lines meet.* For, let us imagine a series of right lines passing in one plane through a common point A. The locus of their most distant points will then, by the last proposition be a complete right line β. Then, suppose this plane to revolve round a Euclidean right line lying in it at the point A. The locus β will then revolve round the point A' in a plane, and will therefore describe a complete plane.

We have here a partially independent proof of proposition II, since the locus in question must be alike in all its parts. The basis of this second proof is proposition IV which rests on the basis that the most distant region of a revolving line describes a Euclidean plane.

VI. *Conversely, all right lines perpendicular to the same complete plane meet in a point at the distance D on each side of the plane.* This point may be called the pole of the plane, and the plane itself may be called the polar plane of the point. The position of a complete plane in space is completely determined by that of its pole, and vice versa. The poles of all planes passing through a point lie in the polar plane of that point.

VII. *For every complete right line, there is a conjugate complete right line such that every point of the one is at distance D from every point of the other.* A line may be changed into its conjugate by two rotations of 90° each around a pair of opposite points.

The three last propositions may be combined as follows. If we call one locus polar to another when every point of the one is at distance D from every point of the other, then, the polar of a point will be a plane, that of a right line will be another right line, and that of a plane will be a point. And, every locus will be completely determined by its polar.

VIII. *Any two planes in space have, as a common perpendicular, the right line joining their poles, and intersect each other in the conjugate to that right line.*

IX. *If a system of right lines pass through a point, their conjugates will be in the polar plane of that point. If they also be in the same plane, the conjugates will all pass through the pole of that plane.*

X. From postulate 3 it may be deduced that the relation between the sides, a, b, and c of a plane triangle in curved space, and their opposite angles A, B, C, will be the same as in a Euclidean spherical triangle of which the corresponding sides are $\frac{a\pi}{2D}$, $\frac{b\pi}{2D}$, and $\frac{c\pi}{2D}$.*

*To prove this, in a rectilineal triangle of which a, b, and c are the sides, and A, B, and C the opposite angles, let us consider b and C as constant, and a, c, and B as functions of A. To find the differential variations of a, c, and B, we substitute dA for α in Postulate III [i.e.,

That is, the relation in question is expressed by the formulae

$$\sin\frac{a\pi}{2D} : \sin\frac{b\pi}{2D} : \sin\frac{c\pi}{2D} = \sin A : \sin B : \sin C. \qquad [2]$$

From this it follows that the right line is a minimum distance between any two points whether we follow it in one direction or the other, that is, whether we consider it as greater or less than D. For, let A and B be the two points, and P the middle point of a line joining AB, lying near the straight line AB. Since $AP = PB < D$, it is evident that the shortest line joining AB and passing through P is composed of the two right lines $AP + PB$. But, by the formulae of spherical trigonometry we have $AB < AP + PB$, so that AB is a minimum line so long as its product by $\frac{\pi}{2D}$ is less than π, that is, so long as it is less than $2D$.

---

postulate 3, eq. [1] above]: we then find[4]

$$\frac{da}{dA} = \frac{2D}{\pi\sin B}\sin\frac{c\pi}{2D}, \qquad [1a]$$

$$\frac{dc}{dA} = \frac{2D}{\pi\tan B}\sin\frac{c\pi}{2D}, \qquad [2a]$$

$$\frac{dB}{dA} = -\cos\frac{c\pi}{2D}. \qquad [3a]$$

The integrals of these equations may be expressed in the form

$$\sin\frac{c\pi}{2D}\sin B = C_1, \qquad [4a]$$

$$\cos\frac{c\pi}{2D} = \sqrt{1-C_1^2}\cos\frac{\pi}{2D}(a-C_2), \qquad [5a]$$

$$\cos B = \sqrt{1-C_1^2}\sin(A-C_3). \qquad [6a]$$

C_1, C_2, and C_3 being arbitrary constants. These constants must be so taken that we shall have simultaneously

$$A = 0, \qquad [7a]$$

$$c = b, \qquad [8a]$$

$$a = 0, \qquad [9a]$$

$$B = 2\pi C \qquad [10a]$$

which conditions give

$$C_1 = \sin C\sin\frac{b\pi}{2D}, \qquad [11a]$$

$$\sqrt{1-C_1^2}\cos\frac{\pi C_2}{2D} = \cos\frac{b\pi}{2D}, \qquad [12a]$$

$$\sqrt{1-C_1^2}\sin C_3 = \cos C. \qquad [13a]$$

When the values of the arbitrary constants derived from these equations are substituted in the integrals, we have the fundamental equations of spherical trigonometry.

XI. *Space is finite, and its total volume admits of being definitely expressed by a number of Euclidean solid units which is a function of D.* We may conceive space as filled in the following way. If a crowd of beings should proceed to form a sphere of matter by building out on all sides from a common centre, they themselves living on the constantly growing surface, then, just before the sphere attained the radius D each being would see those who were diametrically opposite directly above him, so that, in each region, the only vacant space left would be contained between two Euclidean planes separated by the distance $2(D - r)$, r being the radius of the solid sphere. If the building should be continual until these two surfaces met at every point, all space would be filled.

XII. The third postulate affords us the means of readily determining the elementary relations of circular and spherical figures in space. We see at once by the equation

$$ds = \frac{2D}{\pi} \sin \frac{r\pi}{2D} d\alpha \qquad [3]$$

that the circumference of a circle of radius r is $4D \sin \frac{r\pi}{2D}$.[5] When $r = D$ the circle will form two complete right lines, as it should, because the two ends of every diameter then meet. For the area of a circle of radius r we readily find the expression[6]

$$Area = \frac{16D^2}{\pi} \sin^2 \frac{r\pi}{4D}. \qquad [4]$$

The area of a complete plane, counting both sides of the surface, is found by putting $r = 2D$, and is therefore $\frac{16D^2}{\pi}$. This is, in fact, easily found to be the surface generated by a complete right line revolving through 360°. The reason for considering the complete plane as a double surface will be seen presently.

The surface of a sphere of radius r is[7]

$$Surface = \frac{16D^2}{\pi} \sin^2 \frac{r\pi}{2D}, \qquad [5]$$

and its volume[8]

$$\frac{8D^2}{\pi}(r - \frac{D}{\pi} \sin \frac{r\pi}{D}). \qquad [6]$$

The total volume of space will be found by putting $r = D$, which will give

$$Total\ volume = \frac{8D^3}{\pi}. \qquad [7]$$

XIII. *The two sides of a complete plane are not distinct, as in a Euclidean surface.* If we draw a complete straight line on one side of a plane, it will, at the point of completion, be found on the other side, and must be completed a second time if it is to be closed without intersecting the plane. It will, in fact, be a complete circle of radius D. (prop. XII.) If, in the case supposed in XI, just before space is filled, a being should travel to distance $2D$, he would, on his return, find himself on the opposite surface to that on which he started, and would have to repeat his journey in order to return to his original position without leaving the surface. In this property we find a certain amount of reason for considering the complete plane as a double surface.

XIV. The following proposition is intimately connected with the preceding one. If, moving along a right line, we erect an indefinite series of perpendiculars, each in the same Euclidean plane with the one which precedes it, then, on completing the line and returning to our starting point, the perpendiculars will be found pointing in a direction the opposite of that with which we started.

It may be remarked that the law of curvature here supposed does not seem to coincide with one of the conclusions of *Riemann*. The latter says: "If one prolonged the initial directions lying in a surface direction into shortest lines, one would obtain an unbounded surface with constant positive curvature, and thus a surface which in a flat triply extended manifold would take the form of a sphere, and consequently be finite."[9] If, by this is meant that if the triply extended curved space became plane space, the complete plane would become a sphere, a discussion of the proposition would be too long to be entered upon here. I cite it only to remark that the complete plane described in the present paper must by no means be confounded with a sphere from which it differs in several very essential characteristics.

α. It has no diameter; a straight line, whether normal to it or not, only intersects it in a single point.

β. The shortest line connecting any two points of it lies upon it.

γ. The locus of the most distant point upon it is not a point, but a right line.

In the same way, the complete right line does not possess the properties of a circle. It does not intersect its normal plane at more than a single point; the most distant point upon it is, on the contrary, at greatest possible distance from the normal plane.

It may be also remarked that there is nothing within our experience which will justify a denial of the possibility that the space in which we find ourselves may be curved in the manner here supposed. It might be claimed that the distance of the farthest visible star is but a small fraction of the greatest distance D, but nothing more. The subjective impossibility of conceiving of the relation of the most distant points in such a space does not render its existence incredible. In fact our difficulty is not unlike that which must have been felt by the first man to whom the idea of the sphericity of the earth was suggested in conceiving how, by travelling in a constant direction, he could return to the point from which he started without, during his journey, finding any sensible change in the direction of gravity.

Notes

[This follows the text of Newcomb 1877, with equation numbers added in square brackets and the spelling of "Euclidean" is regularized. For context about Newcomb and Einstein's remarks on him, see the Introduction.]

1. [Note that Newcomb calls a straight line a "right line."]

2. [The reason Newcomb assumes this form is that when $r = 2D$, then $s = \frac{2\alpha D}{\pi}\sin\pi = 0$, as required by his postulate 2. Notice also that when $r \ll 2D$, $s = \frac{2\alpha D}{\pi}\frac{r\pi}{2D} \approx \alpha r$ (using the approximation $\sin s \approx x$, if $x \ll 1$), in accord with the usual relation between arc length s angle α, and radius r, $s = \alpha r$, showing that for distances small compared to $2D$, $s \propto r$.]

3. [Taking the derivative of his expression for s gives $\frac{ds}{dr} = \frac{2\alpha D}{\pi}\frac{\pi}{2D}\cos\frac{r\pi}{2D} = \alpha\cos\frac{r\pi}{2D} \approx 0$ if $r \ll 2D/\pi$ because $\cos x \approx 0$ if $x \ll 1$.]

4. [This diagram will help clarify the geometry:

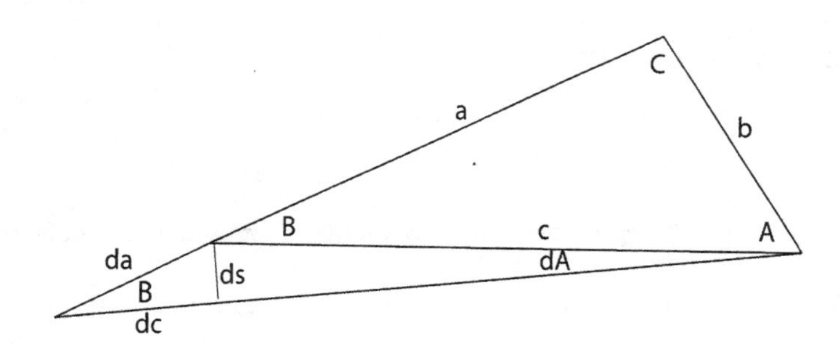

Following postulate 3 (correcting Newcomb's mention of "Postulate III"), we have used c as the distance r, swung through the infinitesimal angle $\alpha = dA$. Then from postulate 3, $ds = \frac{2dAD}{\pi}\sin\frac{c\pi}{2D}$ or $ds/dA = \frac{2D}{\pi}\sin\frac{c\pi}{2D}$. From the diagram, $\sin B \to ds/da$, in the limit of infinitesimal dA. Then $ds = ds/\sin B$ and substituting the value we just derived for ds yields [1a]. For [2a], we use the limiting formula $\tan B \to ds/dc$ and the above expression for ds. For [3a], note that $\sin B \to ds/da$ and take the derivative of both sides with respect to A: $\cos B\frac{dB}{dA} = \frac{d}{dA}(\frac{ds}{da}) = \frac{d}{da}(\frac{ds}{dA})$, reversing the order in which the derivatives are taken. Then we use the above expression for ds/dA, giving $\cos B\frac{dB}{dA} = \frac{dc}{da}\cos\frac{c\pi}{2D}$, and note from the diagram that $dc/da = -\cos B$, because positive dA gives a negative change in B, in terms of the choice of signs in the diagram. Then [3A] follows.

The integrals [4a–6a] can be verified by direct substitution. For instance, using [4a], calculate $\sqrt{1 - C_1^2}$ and then calculate $d/dA(\sqrt{1 - C_1^2})$. After some straightforward use of the chain rule and some algebra, the answer is zero, showing that $\sqrt{1 - C_1^2}$ and hence also C_1 is indeed a constant, as claimed. Similar steps verify [5a–6a]. Since these constants are arbitrary, Newcomb has the freedom to make the choices [7a–10a]. Substitution of the resulting expressions [11a–13a] with the values given in [4a–6a] then yields equation [2] in the main text, the "law of sines" in spherical trigonometry, as summarized in Gauss 2005, 123, and Klaf 2005.]

5. [This follows because for a complete circle $d\alpha \to 2\pi$.]

6. [Write the circumference of a circle $4D\sin\frac{r\pi}{2D} = 4D(2\sin\frac{r\pi}{4D}\cos\frac{r\pi}{4D})$, using the identity that $\sin 2x = 2\sin x\cos x$. Then the area of the circle will be given by the sum of circumferential rings

$$\int_0^r 4D\left(2\sin\frac{r\pi}{4D}\cos\frac{r\pi}{4D}\right)\,dr = 4D(4D/\pi)\sin^2\frac{r\pi}{4D} = \frac{16D^2}{\pi}\sin^2\frac{r\pi}{4D},$$

Newcomb's eq. [4].]

7. [To find the surface area of a sphere of radius r, take the circumference of a circle of the same radius and then rotate that circle one complete revolution around an arbitrary axis, so that the circle sweeps out infinitesimally the an area composed of its circumference times ds, the arc length swept through, and the whole surface area is then

$$\int_0^{2\pi} 4D\sin\frac{r\pi}{2D}\,ds = \int_0^{2\pi} 4D(\frac{2D}{\pi})\sin\frac{r\pi}{2D}\sin\frac{r\pi}{2D}d\alpha,$$

using eq. [3]. The integral over $d\alpha$ gives simply 2π and the whole result is Newcomb's [5].]

8. [Now to find the volume of the sphere, we integrate its surface area

$$\int_0^r \frac{16D^2}{\pi}\sin^2\frac{r\pi}{2D}\,dr = \frac{16D^2}{2\pi}\int_0^r (1-\cos\frac{r\pi}{D})\,dr,$$

using the identity that $\sin^2 x = (1-\cos 2x)/2$. Then eq. [6] follows directly from this integral.]

9. [Newcomb quotes directly the German, for which I have put the translation used above, 32.]

Non-Euclidean Geometries (1891)

Henri Poincaré

Every conclusion assumes premises. These premises themselves are either self-evident and need no demonstration or can only be established as resting on other propositions. Because we cannot go back in this way to infinity, every deductive science, and geometry in particular, must rest upon a certain number of indemonstrable axioms. All geometry textbooks therefore begin with the enunciation of these axioms. But there is a distinction to be made between them; some (for example, "Things which are equal to the same thing are equ to one another") are not geometrical propositions but propositions of analysis. I consider these as analytic a priori judgments and will not discuss them here.[1]

But I must insist on other axioms which are special to geometry. Of these most textbooks explicitly state three:

(1) Only one straight line can pass through two points.

(2) A straight line is the shortest distance between two points.

(3) Through one point, only one parallel can be drawn to a given straight line.

Though we generally dispense with proving the second of these axioms, it would be possible to deduce it from the other two and from those much more numerous axioms that are implicitly admitted without stating them, as I shall explain later.

For a long time, a proof of the third axiom, known as the parallel postulate, was sought in vain. The amount of effort that was spent in pursuit of this chimerical hope is truly beyond imagination. Finally, at the beginning of the century, and almost simultaneously, two scientists, a Russian and a Hungarian, Lobachevsky and Bolyai, showed irrefutably that this proof is impossible. They have nearly rid us of inventors of geometries without the parallel postulate; since then, the Académie des Sciences only receives about one or two new demonstrations a year.

But the question was not exhausted, and it was not long before a great ste was taken by the celebrated memoir of Riemann, entitled "On the Hypothese That Lie at the Foundation of Geometry." This little work inspired most the recent work that I shall discuss, among which I cite those of Beltrami a Helmholtz.[2]

The geometry of Lobachevsky.—If it were possible to deduce the para postulate from other axioms, it is evident that by rejecting that postulate retaining the other axioms we should be led to contradictory consequences would be, therefore, impossible to found a coherent geometry on those prem

Now, this is precisely what Lobachevsky has done. He assumes at the outset that *through a point, several parallels may be drawn to a given straight line*, and he retains all the other axioms of Euclid. From these hypotheses he deduces a series of theorems between which it is impossible to find any contradiction, and he constructs a geometry whose impeccable logic yields nothing to Euclidean geometry.

Certainly, the theorems are very different from those to which we are accustomed and at first they are a little disconcerting.

For instance, the sum of the angles of a triangle is always less than two right angles, and the difference between that sum and two right angles is proportional to the area of the triangle.

It is impossible to construct a figure similar to a given figure but of different dimensions.

If the circumference of a circle be divided into n equal parts and tangents drawn at the points of division, these n tangents will form a polygon, provided that the radius of the circle be small enough; but if the radius is sufficiently large, they will never meet.

It is useless to multiply these examples; Lobachevsky's propositions have no relation to those of Euclid, but they are no less logically interconnected.

The geometry of Riemann.—Let us imagine a world peopled only with beings of no thickness and let us suppose these "infinitely flat" animals are all in the same plane, from which they cannot emerge. Let us further admit that this world is sufficiently distant from other worlds to be removed from their influence. While we are making these hypotheses, it will not cost us much more to endow these beings with rationality and believe them capable of doing geometry. In that case, they will certainly attribute to space only two dimensions.[3]

But now let us suppose that these imaginary animals, while remaining with- thickness, have the form of a spherical, and not of a plane figure, and are all e same sphere, from which they cannot escape. What geometry will they uct? In the first place, it is clear that they will attribute to space only mensions. The straight line to them will be the shortest distance from nt on the sphere to anotherthat is to say, an arc of a great circle. In a eir geometry will be spherical geometry.

they call space will be the sphere on which they are confined and on ur all the phenomena with which they are acquainted. Their space re be *unbounded* because on a sphere one may always walk forward r being stopped, and yet it will be *finite*; the end will never be found, plete circuit can be made.

nann's geometry is spherical geometry extended to three dimen- struct it, the German mathematician had first of all to throw only the parallel postulate, but also the first axiom: *only one rough two points.*

through two given points, we can *in general* draw only one h, as we have just seen, would play for our imaginary beings the role of a straight line. But there is one exception: If the two given points are diametrically opposite, an infinite number of great circles can be drawn through them.

Similarly, in Riemann's geometry through two points only one straight line can in general be drawn, but there are exceptional cases in which through two points an infinite number of straight lines can be drawn.

There is a kind of opposition between the geometries of Riemann and Lobachevsky.

Thus, the sum of the angles of a triangle is equal to two right angles in Euclid's geometry, less than two right angles in that of Lobachevsky, and greater than two right angles in that of Riemann.

The number of parallel lines that can be drawn through a given point to a given line is one in Euclid's geometry, none in Riemann's, and an infinite number in the geometry of Lobachevsky. Let us add that Riemann's space is finite, though unbounded in the sense we gave to these words.

Surfaces of constant curvature.—One objection, however, remains possible. The theorems of Lobachevsky and Riemann present no contradiction, but however numerous may be the other consequences these geometers deduced from their hypotheses, they had to stop before they exhausted them all, for the number would be infinite; who can say whether if they carried their deductions further they would not have eventually reached some contradiction?

This difficulty does not exist for Riemann's geometry, provided it is limited to two dimensions. As we have seen, the two-dimensional geometry of Riemann, in fact, does not differ from spherical geometry, which is only a branch of ordinary geometry, and is therefore outside of this discussion.

Beltrami, by showing that Lobachevsky's two-dimensional geometry was only a branch of ordinary geometry, has equally refuted the objection in this connection.

This is the course of his argument: Let us consider any figure whatever on a surface. Imagine this figure to be traced on a flexible and inextensible canvas applied to the surface, in such a way that when the canvas is displaced and deformed the different lines of the figure change their form without changing their length. As a rule, this flexible and inextensible figure cannot be displaced without leaving the surface, but there are certain particular surfaces for which such a movement would be possible; they are surfaces of constant curvature.

If we resume the comparison that we made just now and imagine beings without thickness living on one of these surfaces, they will think it possible that a figure could move so that all its lines would remain of a constant length. Such a movement would appear absurd, on the other hand, to animals without thickness living on a surface of variable curvature.

These surfaces of constant curvature are of two kinds. Some have positive curvature and may be deformed so as to be applied to a sphere. The geometry of these surfaces is therefore reduced to spherical geometry, which is Riemannian.

Others have *negative curvature.* Beltrami has shown that the geometry of these surfaces is identical with that of Lobachevsky. Thus the two-dimensional geometries of Riemann and Lobachevsky are connected with Euclidean geometry.

Interpretation of non-Euclidean geometries.—Thus vanishes the objection as far as two-dimensional geometries are concerned.

It would be easy to extend Beltrami's reasoning to three-dimensional geometries. Minds who do not recoil before space of four dimensions will see no difficulty in it, but such minds are few in number. I prefer, then, to proceed otherwise.

Let us consider a certain plane, which I shall call the fundamental plane, and let us construct a kind of dictionary by making a double series of terms written in two corresponding columns, just as in ordinary dictionaries the words in two languages that have the same signification correspond to one another:

Space......................	The portion of space situated above the fundamental plane
Plane......................	Sphere orthogonally cutting the fundamental plane
Line......................	Circle orthogonally cutting the fundamental plane
Sphere......................	Sphere
Circle......................	Circle
Angle......................	Angle
Distance between two points	Logarithm of the anharmonic ratio of these two points[4] and of the intersection of the fundamental plane with the circle passing through these two points and cutting it orthogonally
etc.......................	etc. ...

Let us then take Lobachevsky's theorems and translate them with the help of this dictionary, as we would translate a German text with the help of a German-French dictionary. *We shall then obtain the theorems of ordinary geometry.*

For instance, Lobachevsky's theorem: "The sum of the angles of a triangle is less than two right angles" may be translated thus: "If a curvilinear triangle has for its sides arcs of circles that if produced would cut orthogonally the fundamental plane, the sum of the angles of this curvilinear triangle will be less than two right angles." Thus, however far the consequences of Lobachevsky's hypotheses are carried, they will never lead to a contradiction. In fact, if two of Lobachevsky's theorems were contradictory, so would be the translations of these two theorems made with the help of our dictionary. But these translations are theorems of ordinary geometry and no one doubts that ordinary geometry is exempt from contradiction. Where does this certainty come from and is it justified? That is a question upon which I cannot enter here but which is very interesting and not insoluble, I think. Nothing, therefore, is left of the objection I formulated above.

But this is not all. Lobachevsky's geometry, being susceptible of a concrete interpretation, ceases to be a fruitless logical exercise and may be applied. I have no time here to deal with these applications, nor with what Klein and I have done by using them to integrate linear equations.

Further, this interpretation is not unique and several dictionaries may be constructed analogous to that above, all of which will enable us by a simple "translation" to tranform Lobachevsky's theorems into theorems of ordinary geometry.

Implicit Axioms.—Are the axioms implicitly enunciated in our textbooks the only foundation of geometry? We may be assured of the contrary when we see that, when they are abandoned one after another, some propositions are still left standing that are common to the geometries of Euclid, Lobachevsky, and Riemann. These propositions must rest on premises that geometers admit without stating them. It is interesting to try to extract them from the classical proofs.

John Stuart Mill asserted that every definition contains an axiom, because by defining we implicitly affirm the existence of the object defined.[5] That is going rather too far; it is rare in mathematics that a definition is given without following it up by the proof of the existence of the object defined, and when this is not done it is generally because the reader can easily supply it. It must not be forgotten that the word "existence" has not the same meaning when it refers to a mathematical entity as when it refers to a material object. A mathematical entity exists provided there is no contradiction implied in its definition, either in itself, or with the propositions previously admitted.

But if Mill's observation cannot be applied to all definitions, it is nonetheless true for some of them. A plane is sometimes defined in the following manner:

The plane is a surface such that the line joining any two points of it lies wholly on that surface.

This definition manifestly hides a new axiom; it is true we might change it, and that would be preferable, but then we should have to state the axiom explicitly.

Other definitions may give rise to no less important reflections, such as, for example, the one about the equality of two figures: two figures are equal when they can be superposed. To superpose them, one of them must be displaced until it coincides with the other. But how must it be displaced? If we were to ask, no doubt we should be told that it ought to be done without deforming it, as a rigid, invariant solid body is displaced. The vicious circle would then be evident.

As a matter of fact, this definition defines nothing: it has no meaning for a being living in a world in which there are only fluids. If it seems clear to us, it is because we are accustomed to the properties of natural solids which do not much differ from those of the ideal solids, all of whose dimensions remain invariant when displaced.

However imperfect it may be, this definition implies an axiom.

The possibility of the motion of an invariant figure is not a self-evident truth or at least is only an aspect of the parallel postulate and not as an analytic a priori judgment would be.

Moreover, in studying the definitions and the proofs of geometry, we see that we are compelled to admit without proof not only the possibility of this motion, but also some of its properties.

This first arises in the definition of the straight line. Many defective definitions have been given, but the true one is that which is understood in all the proofs in which the straight line intervenes.

"It might happen that the motion of an invariable figure may be such that all the points of a line belonging to the figure are motionless, while all the points situated outside that line are in motion. Such a line would be called a straight line." In this statement, we have deliberately separated the definition from the axiom it implies.

Many proofs, such as those of the cases of the equality of triangles, of the possibility of drawing a perpendicular from a point to a straight line, assume propositions that have not been stated, because they necessarily imply that it is possible to move a figure in space in a certain way.

The fourth geometry.—Among these implicit axioms there is one which seems to me to deserve some attention, not only because it gave the occasion to a recent discussion* but because in abandoning it, we can construct a fourth geometry as coherent as those of Euclid, Lobachevsky, and Riemann.

To prove that we can always draw a perpendicular from a point A on a straight line AB, consider a straight line AC movable about the point A, initially identical with the fixed straight line AB, and then turn it about the point A until it lies in the extension of line AB.

Thus we assume two propositions, first, that such a rotation is possible, and then that it may continue until the two lines lie along the same extended line.

If we admit the first and reject the second, we are led to a series of theorems even stranger than those of Lobachevsky and Riemann, but equally free from contradiction.

I shall give only one of these theorems, and I shall not choose the least remarkable of them. *A real straight line may be perpendicular to itself.*

Lie's Theorem.—The number of axioms implicitly introduced into classical proofs is greater than necessary, and it would be interesting to reduce them to a minimum. We may ask, in the first place, whether this reduction is possible, if the number of necessary axioms and that of imaginable geometries is not infinite.

A theorem due to Sophus Lie is of weighty importance in this discussion. It may be stated as follows:

Let us assume the following premises: (1) space has n dimensions; (2) the movement of an invariant figure is possible; (3) p conditions are necessary to determine the position of this figure in space.

---

*See Renouvier, Léchalas, Calinon, *Revue Philosophique*, June, 1889, *Critique Philosophique*, 30 September and 30 November 1889; *Revue Philosophique*, 1890, p. 158, in particular the discussion on the "postulate of perpendicularity."

The number of geometries compatible with these premises will be limited.

I can even add that if n is given, an upper bound can be assigned to p.

If, therefore, the possibility of the movement is granted, we can only invent a finite (and even rather restricted) number of three-dimensional geometries.

Riemann's geometries.—Nonetheless, this result seems contradicted by Riemann, for that scientist constructs an infinite number of different geometries, that to which his name is usually attached being only a particular case of them.

Everything depends, he says, on the manner in which the length of a curve is defined. Now there is an infinite number of ways of defining this length, and each of them may be the starting-point of a new geometry.

That is perfectly true, but most of these definitions are incompatible with the movement of a invariant figure such as we assume to be possible in Lie's Theorem. These Riemannian geometries, so interesting on various grounds, can never be, therefore, purely analytical, and would not lend themselves to proofs analogous to those of Euclid.

On the nature of axioms.—Most mathematicians regard Lobachevsky's geometry only as a mere logical curiosity; some of them have, however, gone further. Because several geometries are possible, is it certain that our geometry is the one that is true? Experience no doubt teaches us that the sum of the angles of a triangle is equal to two right angles, but this is because the triangles we deal with are too small. According to Lobachevsky, the difference is proportional to the area of the triangle; will not this become discernible when we operate on much larger triangles and when our measurements become more accurate? Euclid's geometry would thus be a provisonal geometry.

Now, to discuss this view we must first of all ask ourselves what is the nature of geometrical axioms?

Are they synthetic a priori judgments, as Kant affirmed?

They would then be imposed upon us with such force that we could not conceive of the contrary proposition, nor build upon it a theoretical edifice. There would be no non-Euclidean geometry.

To convince ourselves of this, let us take a true synthetic a priori judgment, such as the following:

If we have an infinite series of positive integers, all different, there will always be one that is smaller than all the others.

Or this other statement, which is equivalent:

If a theorem is true for the number 1 and for $n+1$, provided it is true of n, it will be true for all positive integers.

Let us next try to get rid of this and by denying this proposition construct a false arithmetic analogous to non-Euclidean geometry. We will not succeed; we will even be tempted at the outset to look upon these judgments as analytic.

Besides, to take up again our fiction of animals without thickness, we can scarcely admit that these beings, if their minds are like ours, would adopt Euclidean geometry, which would be contradicted by all their experience.

Ought we, then, to conclude that the axioms of geometry are experimental truths? But we do not experiment with ideal lines or circles; we only can use material objects. On what, therefore, do the experiences serving as a foundation for geometry rest? The answer is easy.

We have seen above that we constantly reason as if geometrical figures behaved like solids. What geometry borrows from experience would therefore be the properties of these bodies.

But an insurmountable difficulty remains. If geometry were an experimental science, it would not be an exact science but subject to continual revision. But what am I saying? Geometry would from today on be proved to be erroneous, because we know that no rigorously rigid solid exists.

Geometrical axioms are therefore neither synthetic a priori judgments nor experimental facts.

They are *conventions*. Our choice among all possible conventions is *guided* by experimental facts but remains *free* and is only limited by the necessity of avoiding every contradiction. Thus, postulates may remain *rigorously* true even when the experimental laws that have determined their adoption are only approximate.

In other words, *the axioms of geometry* (I do not speak of those of arithmetic) *are only disguised definitions.*

What, then, are we to think of the question: Is Euclidean geometry true?

It has no meaning.

We might as well ask if the metric system is true and the old weights and measures are false, if Cartesian coordinates are true and polar coordinates false. One geometry cannot be more true than another, only more convenient.

Now Euclidean geometry is, and will remain, the most convenient

1) because it is the simplest and not only because of our mental habits or because of whatever kind of direct intuition that we might have of Euclidean space. It is the simplest in itself, just as a polynomial of the first degree is simpler than a polynomial of the second degree.

2) because it agrees sufficiently with the properties of natural solids, those bodies we approach with our limbs and our eyes and with which we make our measuring instruments.

Geometry and astronomy.[6]—We have also asked the question in another way. If Lobachevskian geometry is true, the parallax of a very distant star would be finite; if Riemannian geometry is true, the parallax will be negative. These results seem accessible to experience and we hope that astronomical observations can allow us to decide between the three geometries.

But what we call a straight line in astronomy is simply the trajectory of a light ray. If then, on the contrary, we came to discover negative parallaxes or to demonstrate that all parallaxes are greater than a certain limit, we would have to choose between two conclusions: we must renounce Euclidean geometry or at least modify the laws of optics and admit that light does not propagate strictly in a straight line.

It is useless to add that everybody considers this solution as more advantageous.

Euclidean geometry therefore has nothing to fear from new experiences.

Let me conclude with a little paradox:

Beings whose minds are made like ours and have the same senses as us but who have not received any previous education could receive from the external world suitably chosen impressions that can lead to construing a geometry other than Euclid's and to localizing the phenomena of this external world in a non-Euclidean space or even in a space of four dimensions.

For us, whose education has been made by our real world, if we were transported suddenly into this new world, we would not have difficulties in connecting its phenomena with our Euclidean space.

Those who would dedicate their lives to it could perhaps succeed in representing to themselves the fourth dimension.

I fear that in these last lines I have not been very clear; I could only do that with these new developments, but I have already been too long and those who might be interested in these developments should read Helmholtz.[7]

In my desire to be brief, I have asserted more than I have proved; I hope the reader will pardon me. So much has been written on this subject; there are so many different opinions that the discussion could fill a book.

A letter on non-Euclidean geometries (1892)

To the editor:

Let me respond to the very interesting letter of Mouret[†] not because I wish to have the last word, for I do not pretend to close definitively a discussion that has lasted more than two thousand years, but because this is the occasion for me to present some new observations.

I have sought to emphasize the important role of experience in the genesis of mathematical notions, but at the same time I wished to show that that role is limited. To attain this double goal, the fictions of Riemann and Beltrami, with which I have entertained your readers, can do some service; in fact, they help the imagination to break the habits created by everyday experience that are so inveterate that they seem necessarily imposed on our mind.

Here is one of these fictions that seems quite amusing to me.[8] Let us imagine a sphere S inside which is a medium whose index of refraction and temperature are variable. In this medium are movable objects, but the movements of these objects are sufficiently slow and their specific heats sufficiently low that they immediately come to equilibrium with the temperature of the medium. Further, all these objects have the same coefficient of expansion, so that we can define the absolute temperature by the length of any one of them. Let R be the radius of the sphere, ρ the distance from a point to the center of the spere. I assume

[†]See the *Revue*. vol. III, p. 39 (1892).

that at this point the absolute temperature is $R^2 - \rho^2$ and the index of refraction is $\frac{1}{R^2-\rho^2}$.

What, then, would intelligent beings think who had never left such a world?

1) Because the dimensions of two small objects transported from one point to another would vary *in the same ratio*, because the coefficient of expansion is the same, these beings would believe that these dimensions are not changed; they would have no idea of what we call change of temperature. No thermometer could show it to them because the expansion of its casing would be the same as that of its thermometric liquid.[9]

2) They would believe that this sphere S is infinite; in fact, they would not be able to reach the surface because as they approached it, they would enter colder and colder regions, would become smaller and smaller without realizing it, and would take smaller and smaller steps.

3) What they call straight lines would be orthogonal circumferences of the sphere S, for three reasons:

1) These would be the trajectories of light rays;

2) In measuring different curves with a meter stick, our imaginary beings would recognize that these circumferences are the shortest path from point to another. In fact, their meter stick would contract or expand when they passed from one region to another and they would not realize it;

3) If a solid body were to turn so that that one of its lines remained fixed, that line would be nothing else than one of these circumferences. Thus, if a cylinder were to turn slowly around two spindles and were heated on one side, the locus of its points that remained fixed would be a curve, convex on the heated side, and not straight.

As a result, these beings would adopt Lobachevskian geometry.

But I have wandered quite far from the purpose of my letter; these considerations show the importance of experience and consequently confirm what Mouret reproaches me with. I must insist a little on the differences.

Can experience *and experience alone* engender mathematical notions and (without going as far as Mouret to the foundations of equality) *alone* give us the notion of mathematical continuity? To show the need of doubt, it suffices to reflect on the profound difference that separates physical continuity from mathematical. Consider a sensation that grows gradually; it seems to have something parallel to the continuity of geometers. Fechner has also sought a mathematical expression between sensation and excitation, but on what experiences has he established his celebrated law?[10] We cannot distinguish a weight A of 10 grams from a weight B of 11 grams, nor from that of a weight C of 12 grams, but we distinguish weight A from weight C. Translating these experiences into equations *without distorting them* yields

$$A = B, \quad B = C, \quad A < C.$$

This is the formula of physical continuity, while that of mathematical continuity is

$$A < B < C.$$

But Mouret goes much further in his remarkable letter in the *Revue philosophique*;‡ he attacks the primordial notion of equality that he wishes to derive from experience. I approve much of this article, above all the thought that the idea of space is not a simple idea and that all mathematical ideas resolve themselves into categories of relation, resemblance, difference, and individuality. I took much interest in reading his arguments, whose variety I admired, but I cannot keep from recalling that the most distinctive ones were already in "Counting and Measuring" of Helmholtz, *only different in the conclusions.*[11] I confess that I cannot decide whether to believe that the proposition that two quantities equal to a third are equal to each other is an experimental fact that more precise experiences might perhaps someday invalidate. I would rather conclude with Helmholtz that we give the name of equality to all that in the external world conforms to the preconceived idea we have of mathematical equality.

H. Poincaré

Notes

[Poincaré 1891, 1892a. Note that, until the final section (as indicated in the notes), all this material also appeared in Poincaré's 1902 book *Science and Hypothesis*, Poincaré 1952, 34–50 (and in translation in Poincaré 1892b). Note that the present translation consistently translates the French word *expérience* as "experience," though in French usage *expérience* can readily connote the English "experiment": for instance, *faire une expérience* means "to do an experiment," with an active sense (as opposed to the passive sense of the English expression "to have an experience"). Many French dictionaries include the adjective *expérimental* and the verb *expérimenter*, but lack the parallel noun form "*expériment*." Accordingly, the reader should bear in mind that sometimes Poincaré's "experiences" may be understood as "experiments," but without excluding the broader connection to "experience" as such, for which Einstein's term will be *Erfahrung*.]

1. [For Kant's distinction between analytic and synthetic, see Kant 1998, 32–33, 130–132.]
2. [Beltrami had shown that Lobachevsky's geometry can be interpreted as that of a negatively curved surface (a pseudosphere); see Stillwell 1996, Bonola 1955, 138–139, 173–175, 234–236, and Gray 1989, 147–154. For Poincaré and Riemann, see Boi 1995b.]
3. [This metaphor, which Helmholtz already used, Poincaré will take up in an interesting new context in his letter given below, 105.]

‡On this subject, see the number of the *Revue générale des Sciences* of 30 December 1891, vol. II, p. 826. [this note by the editor of the *Revue*]

4. [For the definition of the anharmonic ratio, see below 114, note 1. The new concept of "distance" in Poincaré's model is defined by the logarithm of this ratio and thus can make the fundamental line have an infinite "distance" from points that (in ordinary distance terms) are finitely removed from it. For more on this model, see Rosenfeld 1988, 239–246.]

5. [See Mill 1974, 218, 224–236, 240–249, 617–618, and Richards 1988, 34–39.]

6. [This section was not carried over fully in Poincaré's reworking of his 1891 text to the 1902 text of his book, except for a brief mention of the being without any education whose impressions would lead it to choose Euclidean geometry and of the assertion that someday a person who devotes his life to it would be able to represent to himself the fourth dimension (at the beginning of chapter IV of Poincaré 1952, 51).]

7. [This explicit mention of Helmholtz has no parallel in Poincaré 1952.]

8. [The extended metaphor of this letter Poincaré incorporated later in his book, Poincaré 1902, 65–68. The specific thought experiment of a varying temperature field was of great importance to Einstein, as discussed in the following note.]

9. [Note Poincaré's assumption that all substances have the same thermal coefficient of expansion (in the context of this thought experiment). Einstein later rejected this condition because of its arbitrariness and even more its universality, requiring an unexplained physical uniformity of every substance. Instead, he thought that geometry gave a far more natural explanation of such uniformity (also seen in the universal equality of inertial and gravitational mass, as in the tower of Pisa experiment) because all objects would respond to spatial curvature in the same way, regardless of their composition and without necessitating any arbitrary assumptions about their response to externally imposed physical fields. See Einstein 1961, 83–86, to see how he uses and replies to this thought experiment]

10. [Fechner's Law in psycho-physiology holds that the magnitude of a subjective sensation is proportional to the logarithm of the stimulus intensity. Riemann also refers to Fechner's theory that all things have living soul as well as to his general methodology; see Laugwitz 1999, 281. For Fechner's work on electrodynamics and his relation to Weber and Helmholtz, see Wise 1981, 278, 283–287.]

11. [This essay is available in translation in Helmholtz 1977, 72–114, and Ewald 1996, 2:727–752.]

The Most Recent Researches in Non-Euclidean Geometry (1893)

Felix Klein

My remarks today will be confined to the progress of non-Euclidean geometry during the last few years. Before reporting on these latest developments, however, I must briefly summarize what may be regarded as the general state of opinion among mathematicians in this field. There are three points of view from which non-Euclidean geometry has been considered.

(1) First we have the point of view of elementary geometry, of which Lobachevsky and Bolyai themselves are representatives. Both begin with simple geometrical constructions, proceeding just like Euclid, except that they substitute another axiom for the parallel postulate. Thus they build up a system of non-Euclidean geometry in which the length of the line is infinite, and the "measure of curvature" (to anticipate a term not used by them) is negative. It is, of course, possible by a similar process to obtain the geometry with a positive measure of curvature, first suggested by Riemann; it is only necessary to formulate the axioms so as to make the length of a line finite, whereby the existence of parallels is made impossible.

(2) From the point of view of projective geometry, we begin by establishing the system of projective geometry in the sense of von Staudt, introducing projective coordinates, so that straight lines and planes are given by *linear* equations.[1] Cayley's theory of projective measurement leads then directly to the three possible cases of non-Euclidean geometry: hyperbolic, parabolic, and elliptic, according as the measure of curvature k is $< 0, = 0$ or > 0. It is here, of course, essential to adopt the system of von Staudt and not that of Steiner, since the latter defines the anharmonic ratio by means of distances of points, and not by pure projective constructions.

(3) Finally, we have the point of view of Riemann and Helmholtz. Riemann starts with the idea of the element of distance ds, which he assumes to be expressible in the form

$$ds = \sqrt{\sum a_{ik} dx_i dx_k}. \qquad [1]$$

Helmholtz, in trying to find a reason for this assumption, considers the motions of a rigid body in space, and derives from these the necessity of giving to ds the form indicated. On the other hand, Riemann introduces the fundamental notion of the *measure of curvature of space.*

The idea of a measure of curvature for the case of two variables, i.e. for a surface in a three-dimensional space, is due to Gauss, who showed that this is an intrinsic characteristic of the surface quite independent of the higher space in which the surface happens to be situated. This point has given rise to a misunderstanding on the part of many non-Euclidean writers. When Riemann attributes to his space of three dimensions a measure of curvature k, he only wants to say that there exists an invariant of the "form" $ds = \sqrt{\sum a_{ik} dx_i dx_k}$; he does not mean to imply that the three-dimensional space necessarily exists as a curved space in a space of four dimensions. Similarly, the illustration of a space of constant positive measure of curvature by the familiar example of the sphere is somewhat misleading. Owing to the fact that on the sphere the geodesic lines (great circles) issuing from any point all meet again in another definite point, antipodal, so to speak, to the original point, the existence of such an antipodal point has sometimes been regarded as a necessary consequence of the assumption of a constant positive curvature. The projective theory of non-Euclidean space shows immediately that the existence of an antipodal point, though compatible with the nature of an elliptic space, is not necessary, but that two geodesic lines in such a space may intersect in one point if at all.*

I call attention to these details in order to show that there is some advantage in adopting the second of the three points of view characterized above, although the third is at least equally important. Indeed, our ideas of space come to us through the senses of vision and motion, the "optical properties" of space forming one source, while the "mechanical properties" form another; the former corresponds in a general way to the projective properties, the latter to those discussed by Helmholtz.

As mentioned before, from the point of view of projective geometry, von Staudt's system should be adopted as the basis. It might be argued that von Staudt practically assumes the axiom of parallels (in postulating a one-to-one correspondence between a pencil of lines and a row of points). But I have shown in the *Math. Annalen*† how this apparent difficulty can be overcome by restricting all constructions of von Staudt to a limited portion of space.

I now proceed to give an account of the most recent researches in non-Euclidean geometry made by Lie and myself. Lie published a brief paper on the subject in the *Berichte* of the Saxon Academy (1886), and a more extensive exposition of his views in the same *Berichte* for 1890 and 1891.[2] These papers contain an application of Lie's theory of continuous groups to the problem formulated by Helmholtz. I have the more pleasure in placing before you the results of Lie's investigations as they are not taken into due account in my paper on the foundations of projective geometry in vol. 37 of the *Math. Annalen* (1890),‡ nor in my (lithographed) lectures on non-Euclidean geometry delivered

*This theory has also been developed by Newcomb, in the *Journal für reine und angewandte Mathematik*, vol. 83 (1877), pp. 293–299 [see above, 89–96].

†"Ueber die sogenannte Nicht-Euklidische Geometrie," Math. Annalen, vol. 6 (1873), pp. 112–145 [Klein 1921, 1:254–305, 311–343; see also Klein 1928].

‡"Zur Nicht-Euklidischen Geometrie," pp. 544–572 [Klein 1921, 1:353–383].

at Göttingen in 1889–90; the last two papers of Lie appeared too late to be considered, while the first had somehow escaped my memory.

I must begin by stating the problem of Helmholtz in modern terminology. The motions of three-dimensional space are ∞^6, and form a group, say G_6.[3] This group is known to have an invariant for any two points p, p', namely the distance $\Omega(p, p')$ of these points. But the form of this invariant (and generally the form of the group) in terms of the co-ordinates $x_1, x_2, x_3, y_1, y_2, y_3$ of the points is not known a priori. The question arises whether the group of motions is fully characterized by these two properties so that none but the Euclidean and the two non-Euclidean systems of geometry are possible.

For illustration Helmholtz made use of the analogous case in two dimensions. Here we have a group of ∞^3 motions; the distance is again an invariant; and yet it is possible to construct a group not belonging to any one of our three systems, as follows.[4]

Let z be a complex variable; the substitution characterizing the group of Euclidean geometry can be written in the well-known form

$$z' = e^{i\phi} z + m + in = (\cos\phi + i\sin\phi)z + m + in. \qquad [2]$$

Now modifying this expression by introducing a complex number in the exponent,

$$z' = e^{(\alpha+i)\phi} z + m + in = e^{\alpha\phi}(\cos\phi + i\sin\phi)z + m + in, \qquad [3]$$

we obtain a group of transformations by which a point (in the simple case $m = 0, n = 0$) would not move about the origin in a circle, but in a logarithmic spiral; and yet this is a group G_3 with three variable parameters m, n, ϕ, having an invariant for every two points, just like the original group. Helmholtz concludes, therefore, that a new condition, that of *monodromy*, must be added to determine our group completely.[5]

I now proceed to the work of Lie. First as to the results: Lie has confirmed those of Helmholtz with the single exception that in space of three dimensions the axiom of monodromy is not needed, but that the groups to be considered are fully determined by the other axioms. As regards the proofs, however, Lie has shown that the considerations of Helmholtz must be supplemented. The matter is this. In keeping one point of space fixed, our G_6 will be reduced to a G_3. Now Helmholtz inquires how the differentials of the lines issuing from the fixed point are transformed by this G_3. For this purpose he writes down the formulae

$$\begin{aligned} dx_1' &= a_{11}dx_1 + a_{12}dx_2 + a_{13}dx_3, \\ dx_2' &= a_{21}dx_1 + a_{22}dx_2 + a_{23}dx_3, \\ dx_3' &= a_{31}dx_1 + a_{32}dx_2 + a_{33}dx_3, \end{aligned}$$

and considers the coefficients $a_{11}, a_{12}, \ldots, a_{33}$ as depending on three variable parameters.[6] But Lie remarks that this is not sufficiently general. The linear equations given above represent only the first terms of a power series, and the possibility must be considered that the three parameters of the group may not all be involved in the linear terms. In order to treat all possible cases, the general developments of Lie's theory of groups must be applied, and this is just what Lie does.

Let me now say a few words on my own recent researches in non-Euclidean geometry which will be found in a paper published in the *Math. Annalen*, vol. 37 (1890), p. 544 [Klein 1921, 1:353–383]. Their result is that our ideas as to non-Euclidean space are still very incomplete. Indeed, all the researches of Riemann, Helmholtz, Lie, consider only a portion of space surrounding the origin; they establish the existence of analytic laws in the vicinity of that point. Now this space can of course be continued, and the question is to see what kind of connection of space may result from this continuation. It is found that there are different possibilities, each of the three geometries giving rise to a series of subdivisions.

To understand better what is meant by these varieties of connection, let us compare the geometry on a sphere with that in the sheaf of lines formed by the diameters of the sphere. Considering each diameter as an infinite line or ray passing through the center (not a half-ray issuing from the center), to each line of the sheaf there will correspond two points on the sphere, namely the two points of intersection of the line with the sphere. We have, therefore, a one-to-two correspondence between the lines of the sheaf and the points of the sphere. Let us now take a small area on the sphere; it is clear that the distance of two points contained in this area is equal to the angle of the corresponding lines of the sheaf. Thus the geometry of points on the sphere and the geometry of lines in the sheaf are identical as far as small regions are concerned, both corresponding to the assumption of a constant positive measure of curvature. A difference appears, however, as soon as we consider the whole closed sphere on the one hand and the complete sheaf on the other. Let us take, for instance, two geodesic lines of the sphere, i.e. two great circles, which evidently intersect in two (diametral) points.[7] The corresponding pencils of the sheaf have only one straight line in common.

A second example for this distinction occurs in comparing the geometry of the Euclidean plane with the geometry on a closed cylindrical surface. The latter can be developed in the usual way into a strip of the plane bounded by two parallel lines, as will appear from Fig. 1, the arrows indicating that the opposite points of the edges are coincident on the cylindrical surface. We notice at once the difference: while in the plane all geodesic lines are infinite, on the cylinder there is one geodesic line that is of finite length, and while in the plane two geodesic lines always intersect in one point, if at all, on the cylinder there may be ∞ points of intersection.

This second example was generalized by Clifford in an address before the Bradford meeting of the British Association (1873).[8] In accordance with Clifford's general idea, we may define a closed surface by taking a parallelogram out of an ordinary plane and making the opposite edges correspond point to point as indicated in Fig. 2.

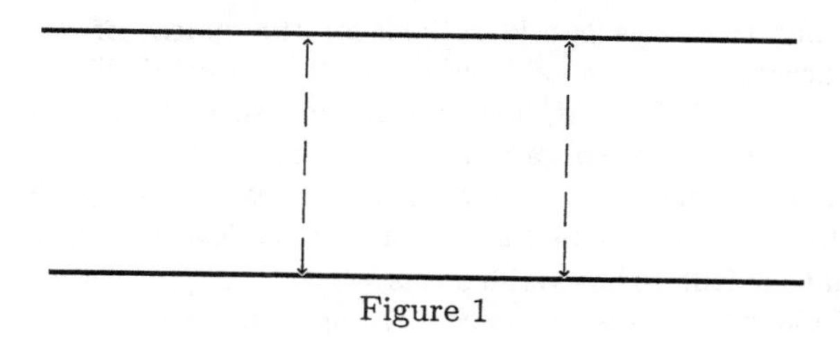

Figure 1

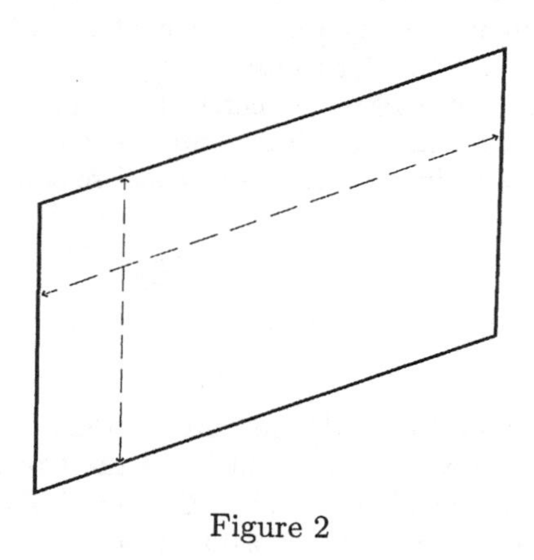

Figure 2

It is not to be understood that the opposite edges should be brought to coincidence by bending the parallelogram (which evidently would be impossible without stretching); but only the logical convention is made that the opposite points should be considered as identical. Here, then, we have a closed manifoldness of the connectivity of an anchor-ring, and everyone will see the great differences that exist here in comparison with the Euclidean plane in everything concerning the lengths and the intersections of geodesic lines, etc.[9]

It is interesting to consider the G_3 of Euclidean motions on this surface. There is no longer any possibility of moving the surface on itself in ∞^3 ways, the closed surface being considered in its totality. But there is no difficulty in moving any small area over the closed surface in ∞^3 ways.

We have thus found, in addition to the Euclidean plane, two other forms of surfaces: the strip between parallels and Clifford's parallelogram. Similarly we have by the side of ordinary Euclidean space three other types with the Euclidean element of arc; one of these results from considering a parallelepiped.

Here I introduce the axiomatic element. There is no way of proving that the whole of space can be moved in itself in ∞^6 ways; all we know is that small portions of space can be moved in space in ∞^6 ways. Hence there exists the possibility that our actual space, the measure of curvature being taken as zero, may correspond to any one of the four cases.[10]

Carrying out the same considerations for the spaces of constant positive measure of curvature, we are led back to the two cases of elliptic and spherical geometry mentioned before. If, however, the measure of curvature be assumed as a negative constant, we obtain an infinite number of cases, corresponding exactly to the configurations considered by Poincaré and myself in the theory of automorphic functions. This I shall not stop to develop here.

I may add that Killing has verified this whole theory.[§] It is evident that from this point of view many assertions concerning space made by previous writers are no longer correct (e.g. that infinity of space is a consequence of zero curvature), so that we are forced to the opinion that our geometrical demonstrations have no absolute objective truth, but are true only for the present state of our knowledge. These demonstrations are always confined within the range of the space-conceptions that are familiar to us; and we can never tell whether an enlarged conception may not lead to further possibilities that would have to be taken into account. From this point of view we are led in geometry to a certain modesty, such as is always in place in the physical sciences.[12]

Notes

[Included as Lecture XL in Klein 1911, delivered in English on September 8, 1893 at Northwestern University "before members of the Congress of Mathematics held in connection with the World's Fair in Chicago." For a helpful discussion in the light of Riemann's lecture, see Norton 1999.]

1. [Projective geometry concerns the properties of figures that are left unchanged when the figures are subject to projective transformations, such as looking at it from a different angle. Euclidean (or "metric") geometry might be regarded, then, as a special case of projective geometry, which Klein and others considered a particularly important way to approach non-Euclidean geometries. Here Klein contrasts the work of Jakob Steiner (1796–1863) with that of K. G. C. von Staudt (1798–1867), who had been a student of Gauss's. Steiner had favored synthetic over analytic methods and treated projective geometry based on metric considerations, using the *cross ratio* or *anharmonic ratio* of four points x_1, x_2, x_3, x_4 defined as $(x_1 - x_3)/(x_1 - x_4) : (x_2 - x_3)/(x_2 - x_4)$, which is invariant under projective transformations. Von Staudt then used "harmonic" sets

[§] "Ueber die Clifford-Klein'schen Raumformen," *Math. Annalen*, vol. 39 (1891), pp. 257–278.[11]

of points (those whose cross ratio is -1) to build up projective geometry without a concept of distance. Arthur Cayley (1821–1895) provided a new metric approach to projective geometry and also considered its generalizations to n dimensions; he also wrote that "projective geometry contains all geometry," cited in Laugwitz 1999, 247. See also Russell 1956, 117–147, Boyer and Merzbach 1991, 537–540, Yaglom 1988, 28–45, and Richards 1979 and 1988, 117–158.]

2. [See also Lie 1967 for a compilation of his results on the Helmholtz-Lie space problem.]

3. [As Klein clarifies below, he is here considering the possible motions of a given point in three-dimensional space with respect to a certain origin. Because both the origin and the given point have each three independent coordinates in which they can move, each of which can take on an infinity of values, then both together have six, so that the group of motions for three-dimensional space as a whole has ∞^6 possible values and hence Klein names it G_6. Note that the invariant is the distance between the two points, $\Omega(p, p')$. This is an example of Klein's Erlangen Program in action, here characterizing Euclidean space by its group of motions. See Scholz 1980, 123–136; see also Norton 1999.]

4. [In the following eq. [2], Klein writes the possible transformations that leave Euclidean geometry invariant as composed of a general translation (e.g., relocation of the origin), characterized by the two quantities m, n, and a general rotation, characterized by the angle ϕ. Since each of these three quantities can taken on an infinity of values, Klein characterizes the group as having ∞^3 motions.]

5. [Recall that monodromy means the symmetry of rotation by 360°, represented in eq. [3] by the (complex) number a in the exponent; $e^{a\phi}$ effectively introduces a rotation as if we were moving out in a logarithmic spiral. Because it seems that this new parameter a would introduce an additional degree of freedom of motion beyond the three already included in the group G_3 (see note 4, above), Helmholtz had thought that he would have to exclude a on the grounds that it would violate the symmetry of monodromy. In the following paragraph, Klein summarizes the results of Lie's more complete arguments, which showed that considerations of monodromy are not necessary to restrict the group G_3.]

6. [These three variable parameters correspond to the three degrees of motion of the movable point.]

7. [By "(diametral) points" Klein seems to mean "diametrically opposite points" or antipodes.]

8. [For the relevant passage of this address by Clifford, see above, 86.]

9. [Klein's examples each differ in what we now would call their topologies (the older term analysis situs was still in use by Cartan, below): though (as Gauss proved) a cylinder and a flat Euclidean plane have the same intrinsic (Gaussian) curvature, their topologies are quite different, shown by the finite geodesic on the cylinder, whereas all geodesics on the plane are infinite. By "anchor-ring" Klein simply means a torus, generated by rotating a circle around a non-intersecting axis in its plane (as in Klein 1963, 46). Klein also uses here as an example the famous Clifford-Klein surface, which locally is everywhere flat,

but all of whose geodesics are finite as a consequence of the identification of the opposite edges, as shown in Fig. 2; for the "Clifford surfaces," see Clifford 1968, 192–193, and Rosenfeld 1988, 298–300. This leads to the general Clifford-Klein problem: to determine all the two-dimensional manifolds of constant curvature that are everywhere regular (as are these two examples). In contrast with fig. 2, which identifies corresponding edges of a parallelogram, a "Klein bottle" identifies *opposite* edges; see Blackett 1967, 40–41, and Lietzmann 1965, 116–118.]

10. [Thus Klein introduces the possibility that the topology of our actual physical space has different possibilities independent of its curvature, even if that curvature is zero.]

11. [For Killing's work, see Hawkins 2000, 111–124.]

12. [Compare this with the final sentence of Riemann's lecture, 33 above. Klein's term "geometry" may here implicitly include what we call topology, so that his argument implies that both the curvature and the topology of space are experimental matters, separately.]

On the Foundations of Geometry (1898)

Henri Poincaré

Although I have already had occasion to set forth my views on the foundations of geometry,* it will not, perhaps, be unprofitable to revert to the question with new and ampler developments, and seek to clear up certain points which the reader may have found obscure. It is with reference to the definition of the point and the determination of the number of dimensions that new light appears to me most needed; but I deem it opportune, nevertheless, to take up the question from the beginning.

Sensible space

Our sensations cannot give us the notion of space. That notion is built up by the mind from elements which pre-exist in it, and external experience is simply the occasion for its exercising this power, or at most a means of determining the best mode of exercising it.

Sensations by themselves have no spatial character.

This is evident in the case of isolated sensations—for example, visual sensations. What could a man see who possessed but a single immovable eye? Different images would be cast upon different points of his retina, but would he be led to classify these images as we do our present retinal sensations?

Suppose images formed at four points A, B, C, D of this immovable retina. What ground would the possessor of this retina have for saying that, for example, the distance AB was equal to the distance CD? We, constituted as we are, have a reason for saying so, because we know that a *slight* movement of the eye is sufficient to bring the image which was at A to C, and the image which was at B to D. But these slight movements of the eye are impossible for our hypothetical man, and if we should ask him whether the distance AB was equal to the distance CD, we should seem to him as ridiculous as would a person appear to us who should ask us whether there was more difference between an olfactory sensation and a visual sensation than between an auditive sensation and a tactual sensation.

But this is not all. Suppose that two points A and B are very near to each other, and that the distance AC is very great. Would our hypothetical man be cognisant of the difference? We perceive it, we who can move our eyes, because

*Both in the *Revue Générale des Sciences* and in the *Revue de Métaphysique et de Morale* [see above, 97–108].

a very slight movement is sufficient to cause an image to pass from A to B. But for him the question whether the distance AB was very small as compared with the distance AC would not only be insoluble, but would be devoid of meaning.

The notion of the contiguity of two points, accordingly, would not exist for our hypothetical man. The rubric, or category, under which he would arrange his sensations, if he arranged them at all, would consequently not be the space of the geometer and would probably not even be continuous, since he could not distinguish small distances from large. And even if it were continuous, it could not, as I have abundantly shown elsewhere, be either homogeneous, isotropic, or tridimensional.

It is needless to repeat for the other senses what I have said for sight. Our sensations differ from one another qualitatively, and they can therefore have no common measure, no more than can the gramme and the metre. Even if we compare only the sensations furnished by the same nerve-fibre, considerable effort of the mind is required to recognize that the sensation of to-day is of the same kind as the sensation of yesterday, but greater or smaller; in other words, to classify sensations according to their character, and then to arrange those of the same kind in a sort of scale, according to their intensity. Such a classification cannot be accomplished without the active intervention of the mind, and it is the object of this intervention to refer our sensations to a sort of rubric or category pre-existing in us.

Is this category to be regarded as a "form of our sensibility"? No, not in the sense that our sensations, individually considered, could not exist without it. It becomes necessary to us only for comparing our sensations, for reasoning upon our sensations. It is therefore rather a form of our understanding.

This, then, is the first category to which our sensations are referred. It can be represented as composed of a large number of scales absolutely independent of one another. Further, it simply enables us to compare sensations of the same kind and not to measure them, to perceive that one sensation is greater than another sensation, but not that it is twice as great or three times as great.

How much such a category differs from the space of the geometer! Shall we say that the geometer admits a category of quite the same kind, where he employs three scales such as the three axes of co-ordinates? But in our category we have not three scales only, but as many as there are nerve-fibres. Further, our scales appear to us as so many separate worlds fundamentally distinct, while the three axes of geometry all fulfil the same office and may be interchanged one for another. In fine, the co-ordinates are susceptible of being measured and not simply of being compared. Let us see, therefore, how we can rise from this rough category which we may call sensible space to geometric space.

The feeling of direction

It is frequently said that certain of our sensations are always accompanied by a peculiar feeling of direction, which gives to them a geometrical character. Such are visual and muscular sensations. Others on the contrary like the sensations of smell and taste are not accompanied by this feeling, and consequently are void of any geometrical character whatever. On this theory the notion of direction would be pre-existent to all visual and muscular sensations and would be the underlying condition of the same.

I am not of this opinion; and let us first ask if the feeling of direction really forms a constituent part of the sensation. I cannot very well see how there can be anything else in the sensation than the sensation itself. And be it further observed that the same sensation may, according to circumstances, excite the feeling of different directions. Whatever be the position of the body, the contraction of the *same* muscle, the biceps of the right arm, for example, will always provoke the *same* muscular sensation; and yet, through being apprised by other concomitant sensations that the position of the body has changed, we also know perfectly well that the direction of the motion has changed.

The feeling of direction, accordingly, is not an integrant part of the sensation, since it can vary without the sensation being varied. All that we can say is that the feeling of direction is associated with certain sensations. But what does this signify? Do we mean by it that the sensation is associated with a certain indescribable something which we can represent to ourselves but which is still not a sensation? No, we mean simply that the various sensations which correspond to the same direction are associated *with one another*, and that one of them calls forth the others in obedience to the ordinary laws of association of ideas. Every association of ideas is a product of habit merely, and it would be necessary for us to discover how the habit was formed.

But we are still far from geometrical space. Our sensations have been classified in a new manner: those which correspond to the same direction are grouped together; those which are isolated and have reference to no direction are not considered. Of the innumerable scales of sensations of which our sensible space was formed some have disappeared, others have been merged into one another. Their number has been diminished.

But the new classification is still not space; it involves no idea of measurement; and, furthermore, the restricted category so reached would not be an isotropic space, that is to say, different directions would not appear to us a fulfilling the same office and as interchangeable with one another. And so this "feeling of direction" far from explaining space would itself stand in need of explanation. But will it help us even towards the explanation we seek? No, because the laws of that association of ideas which we call the feeling of direction are extraordinarily complex. As I explained above, the same muscular sensation may correspond to a host of different directions according to the position of the body which is made known to us by other concomitant sensations. Associations so complex can only be the result of an extremely long process. This, therefore, is not the path which will lead us most quickly to our goal. Therefore we will not regard the feeling of direction as something attained but will revert to the "sensible space" with which we started.

Representation of space

Sensible space has nothing in common with geometrical space. I believe that few persons will be disposed to contest this assertion. It would be possible, perhaps, to refine the category which I set up at the beginning of this article, and to construct something which would more resemble geometrical space. But whatever concession we might make, the space so constructed would be neither infinite, homogeneous, nor isotropic: it could be such only by ceasing to be accessible to our senses.

Seeing that our representations are simply the reproductions of our sensations, therefore we cannot image geometrical space. We cannot represent to ourselves objects in geometrical space, but can merely reason upon them as if they existed in that space.

A painter will struggle in vain to construct an object of three dimensions upon canvas. The image which he traces, like his canvas, will never have more than two. When we endeavor, for example, to represent the sun and the planets in space, the best we can do is to represent the visual sensations which we experience when five or six tiny spheres are set revolving in close proximity.

Geometrical space, therefore, cannot serve as a category for our representations. It is not a form of our sensibility. It can serve us only in our reasonings. It is a form of our understanding.

Displacement and alteration

We at once perceive that our sensations vary, that our impressions are subject to change. The laws of these variations were the cause of our creating geometry and the notion of geometrical space. If our sensations were not variable, there would be no geometry.

But that is not all. Geometry could not have arisen unless we had been led to distribute into two classes the changes which can arise in our impressions. We say, in one case, that our impressions have changed because the objects causing them have undergone some alterations of character, and again that these impressions have changed because the objects have suffered displacement. What is the foundation of this distinction?

A sphere of which one hemisphere is blue and the other red, is rotating before our eyes and shows first a blue hemisphere and then a red hemisphere. Again, a blue liquid contained in a vase suffers a chemical reaction which causes it to turn red. In both cases the impression of blue has given way to the impression of red. Now why is the first of these changes classed among displacements, and the second among alterations? Evidently because in the first case it is sufficient for mc merely to go around the globe to bring myself face to face again with the other hemisphere, and so to receive a second time the impression of blue.

An object is displaced before my eye, and its image which was first formed on the centre of the retina is now brought to the edge of the retina. The sensation which was carried to me by a nerve-fibre proceeding from the centre of the retina is succeeded by another which is carried to me by a fibre proceeding from the edge. These sensations are conducted to me by two different nerves. They ought to appear to me different in character, and if they did not, how could I distinguish them?

Why, then, do I come to conclude that the *same* image has been displaced? Is it because one of these sensations frequently succeeds the other? But similar successions are frequent. These it is that produce all our associations of ideas, and we do not ordinarily conclude that they are due to displacement of an object which is invariable in character.

But what happens in this case is that we can *follow the object with the eye*, and by a displacement of our eye which is generally voluntary and accompanied by muscular sensations, we can bring the image back to the centre of the retina and so *re-establish the primitive sensation.* The following, therefore, is my conclusion.

Among the changes which our impressions undergo, we distinguish two classes:

(1) The first are independent of our will and not accompanied by muscular sensations. These are *external changes* so called.

(2) The others are voluntary and accompanied by muscular sensations. We may call these *internal changes.*

We observe next that in certain cases when an external change has modified our impressions, we can, by voluntarily provoking an internal change, re-establish our primitive impressions. The external change, accordingly, can be *corrected* by an internal change. External changes may consequently be subdivided into the two following classes:

1. Changes which are susceptible of being corrected by an internal change. These are *displacements.*

2. Changes which arc not so susceptible. These are *alterations.* An immovable being would be incapable of making this distinction. *Such a being, therefore, could never create geometry,*—even if his sensations were variable, and even if the objects surrounding him were movable.

Classification of displacements

A sphere of which one hemisphere is blue and the other red, is rotating before me and presents to me first its blue side and then its red side. I regard this external change as a displacement because I can correct it by an internal change, namely, by going around the sphere. Let us repeat the experiment with another sphere, of which one hemisphere is green and the other yellow. The impression of the yellow hemisphere will succeed that of the green, as before that of the red succeeded that of the blue. For the same reason I shall regard this new external change as a displacement.

But this is not all. I also say that these two external changes are due to the same displacement, that is to say, to a rotation. Yet there is no connexion between the impression of the yellow hemisphere and that of the red, any more than there is between that of the blue and that of the green, and I have no reason for saying that the same relation exists between the yellow and the green as exists between the red and the blue. No, I say that these two external changes are due to the same displacement because I have "corrected" them by the same internal change. But how am I to know that the two internal changes by which I corrected first the external change from the blue to the red, then that from the green to the yellow, are to be considered identical? Simply because they have provoked the *same* muscular sensations; and for this it is not necessary for me to know geometry in advance and to represent to myself the movements of my body in geometric space.

Thus several external changes which in themselves have no common relation may be corrected by the same internal change. I collect these into the came class and consider them as the same displacement.

An analogous classification may he made with respect to internal changes. All internal changes arc not capable of correcting an external change. Only those which are may be called displacements. On the other hand the same external change may be corrected by several different internal changes. A person knowing geometry might express this idea by saying that my body can go from the position A to the position B by several different paths. Each of these paths corresponds to a series of muscular sensations; and at present I am cognisant of nothing but these muscular sensations. No two of these series have a common resemblance, and if I consider them nevertheless as representing the *same* displacement, it is because they are capable of correcting the same external change.

The foregoing classification suggests two reflexions:

1. The classification is not a crude datum of experience, because the aforementioned compensation of the two changes, the one internal and the other external, is never exactly realised. It is, therefore, an active operation of the mind, which endeavors to insert the crude results of experience into a pre-existing form, into a category. This operation consists in identifying two changes because they possess a common character, and in spite of their not possessing it exactly. Nevertheless, the very fact of the minds having occasion to perform this operation is due to experience, for experience alone can teach it that the compensation has approximately been effected.

2. The classification further brings us to recognize that two displacements are identical, and it hence results that a displacement can be *repeated* twice or several times. It is this circumstance that introduces number, and that permits measurement where formerly pure quality alone held sway.

Introduction of the notion of group

That we are able to go farther is due to the following fact, the importance of which is cardinal.

It is obvious that if we consider a change A, and cause it to be followed by another change B, we are at liberty to regard the *ensemble* of the two changes

A followed by B as a single change which may be written $A + B$ and may be called the resultant change. (It goes without saying that $A+B$ is not necessarily identical with $B + A$.) The conclusion is then stated that if the two changes A and B are displacements, the change $A + B$ also is a displacement. Mathematicians express this by saying that *the ensemble, or aggregate, of displacements is a group*. If such were not the case there would be no geometry.

But how do we know that the *ensemble* of displacements is a group? Is it by reasoning a priori? Is it by experience? One is tempted to reason a priori and to say: if the external change A is corrected by the internal change A', and the external change B by the internal change B', the resulting external change $A + B$ will be corrected by the resuiting internal change $B' + A'$. Hence this resulting change is by definition a displacement, which is to say that the *ensemble* of displacements forms a group.

But this reasoning is open to several objections. It is obvious that the changes A and A' compensate each other; that is to say, that if these two changes are made in succession, I shall find again my original impressions,—a result which I might write as follows:

$$A + A' = 0.$$

I also see that $B + B' = 0$. These are hypotheses which I made at the outset and which served me in defining the changes A, A', B, and B'. But is it certain that we shall still have $B + B' = O$,—*after* the two changes A and A'? Is it certain that these two changes compensate in such a manner that not only shall I recover my original impressions, but that the changes B and B' shall recover all their original properties, and in particular that of mutual compensation? If we admit this, we may conclude from it that I shall recover my primitive impressions when the four changes follow in the order

$$A, A', B, B';$$

but not that the same will still be the case when they succeed in the order

$$A, B, B', A'.$$

Nor is this all. If two external changes α and α' are regarded as identical on the basis of the convention adopted above, or in other words, are susceptible of being corrected by the same internal change A; if, on the other hand, two other external changes β and β' can be corrected by the same internal change B, and consequently may also be regarded as identical, have we the right to conclude that the two changes $\alpha + \alpha'$ and $\beta + \beta'$ are susceptible of being corrected by the same internal change, and are consequently identical? Such a proposition is in no wise evident, and if it be true it cannot be the result of a priori reasoning.

Accordingly, this set of propositions, which I recapitulate by saying that displacements form a group, is not given us by a priori reasoning. Are they then a result of experience? One is inclined to admit that they are; and yet one has a feeling of real misgiving in so doing. May not more precise experience prove some day that the law above enunciated is only approximate? What, then, will become of geometry? But we may rest assured on this point. Geometry is safe from all revision; no experience, however precise, can overthrow it. If it could have done it, it would have done so long ago. We have long known that all the so-called experimental laws are approximations, and rough approximations at that.

What, then, is to be done? When experience teaches us that a certain phenomenon does not correspond *at all* to these laws, we strike it from the list of displacements. When it teaches us that a certain change obeys them *only approximately*, we consider the change, *by an artificial convention*, as the resultant of two other component changes. The first component is regarded as a displacement *rigorously* satisfying the laws of which I have just spoken, while the second component, which is small, is regarded as a qualitative alteration. Thus we say that natural solids undergo not only great changes of position but also small flexions and small thermal dilatations.

By an external change α we pass, for example, from the *ensemble* of impressions A to the *ensemble* B. We correct this change by a voluntary internal change β and are carried back to the *ensemble* A. A new external change α' causes us to pass again from the *ensemble* A to the *ensemble* B. We ought to expect then that this change α' could in its turn be corrected by another voluntary internal change β which would provoke the same muscular sensations as β and which would call forth again the *ensemble* of impressions A. If experience does not confirm this prediction, we shall not be embarrassed. We say that the change α', although like α it has been the cause of my passing from the *ensemble* A to the *ensemble* B, is nevertheless not identical with the change α. If our prediction is confirmed only approximately we say that the change α' is a displacement identical with the displacement α but accompanied by a slight qualitative alteration.

In fine, these laws are not imposed by nature upon us but are imposed by us upon nature. But if we impose them upon nature, it is because she suffers us to do so. If she offered too much resistance, we should seek in our arsenal for another form which would be more acceptable to her.

Consequences of the existence of the group

This first fact, that displacements form a group, contains in germ a host of important consequences. Space must be homogeneous; that is, all its points are capable of playing the same part. Space must be isotropic; that is, all directions which issue from the same point must play the same part.

If a displacement D transports me from one point to another, or changes my orientation, I must after such displacement D be still capable of the same movements as before the displacement D, and these movements must have preserved their fundamental properties, which permitted me to classify them among displacements. If it were not so, the displacement D followed by another displacement would not be equivalent to a third displacement; in other words, displacements would not form a group.

Thus the new point to which I have been transported plays the same part as that at which I was originally; my new orientation also plays the same part as the old; space is homogeneous and isotropic.

Being homogeneous, it will be unlimited; for a category that is limited cannot be homogeneous, seeing that the boundaries cannot play the same part as the centre. But this does not say that it is infinite; for the sphere is an unbounded surface, and yet it is finite. All these consequences, accordingly, are germinally contained in the fact which we have just discovered. But we are as yet unable to perceive them, because we do not yet know what a direction is or even what a point is.

Properties of the group

We have now to study the properties of the group. These properties are purely formal. They are independent of any quality whatever, and in particular of the qualitative character of the phenomena which constitute the change to which we have given the name displacement. We remarked above that we could regard two changes as representing the same displacement, although the phenomena were quite different in qualitative nature. The properties of this displacement remain the same in the two cases; or rather the only ones which concern us, the only ones which are susceptible of being studied mathematically, are those in which quality is in no wise concerned. A brief digression is necessary here to render my thought comprehensible. What mathematicians call a group is the *ensemble* of a certain number of operations and of all the combinations which can be made of them. In the group which is occupying us our operations are displacements. It sometimes happens that two groups contain operations which are entirely different as to character, and that these operations nevertheless combine according to the same laws. We then say that the two groups are *isomorphic.*

The different permutations of six objects form a group and the properties of this group are independent of the character of the objects. If in place of the six material objects we take six letters, or even the six faces of a cube, we obtain groups which differ as to their component materials, but which are all isomorphic with one another.

The formal properties are those which are common to all isomorphic groups. If I say, for example, that such and such an operation repeated three times is equivalent to such and such an other repeated four times, I have announced a formal property entirely independent of quality. These formal properties are susceptible of being studied mathematically. They should be enunciated, therefore, in *precise* propositions. On the other hand, the experiences which serve to verify them can never be more than approximate; that is to say, the experiences in question can never be the true foundation of these propositions. We have within us, in a potential form, a certain number of models of groups, and experience merely assists us in discovering which of these models departs least from reality.

Continuity

It is observed first that the group is *continuous.* Let us see what this means, and how the fact can be established.

The same displacement can be repeated twice, three times, etc. We obtain thus different displacements which may be regarded as *multiples* of the first. The multiples of the same displacement D form a group; for the succession of two of these multiples is still a multiple of D. Further, all these multiples are interchangeable (a truth which is expressed by saying that the group which they form is a *sheaf*); that is, it is indifferent whether we repeat D first three times and then four times, or first four times and then three times. This is an analytical judgment a priori; an out-and-out tautology. This group of the multiples of D is only a part of the total group. It is what is called a *sub-group.*

Now we soon discover that any displacement whatever can always be divided into two, three, or any number of parts whatever; I mean that we can always find an other displacement which, repeated two, three times will reproduce the given displacement. This divisibility to infinity conducts us naturally to the notion of mathematical continuity; yet things are not so simple as they appear at first sight.

We cannot prove this divisibility to infinity, directly. When a displacement is very small, it is inappreciable for us. When two displacements differ very little, we cannot distinguish them. If a displacement D is extremely small, its consecutive multiples will be indistinguishable. It may happen then that we cannot distinguish $9D$ from $10D$, nor $10D$ from $1lD$, but that we can nevertheless distinguish $9D$ from $11D$. If we wanted to translate these crude facts of experience into a formu1a, we should write

$$9D = 10D, \quad 10D = 11D, \quad 9D < 11D.$$

Such would be the formula of physical continuity. But such a formula is repugnant to reason. It corresponds to none of the models which we carry about in us. We escape the dilemma by an artifice; and for this physical continuity—or, if you prefer, for this sensible continuity, which is presented in a form unacceptable to our minds—we substitute mathematical continuity. Severing our sensations from that something which we call their cause, we assume that the something in question conforms to the model which we carry about in us, and that our sensations deviate from it only in consequence of their crudeness.

The same process recurs every time we apply measurement to the data of the senses; it is notably applicable to the study of displacements. From the point which we have now reached, we can render an account of our sensations in several different ways.

(1) We may suppose that each displacement forms part of a sheaf formed of all the multiples of a certain small displacement far too small to be appreciated by us. We should then have a discontinuous sheaf which would give us the illusion of physical continuity because our gross senses would be unable to distinguish any two consecutive elements of the sheaf.

(2) We may suppose that each displacement forms part of a more complex and richer sheaf. All the displacements of which this sheaf is composed would be interchangeable. Any two of them would be multiples of another smaller displacement which likewise formed part of the sheaf and which might be regarded as their greatest common divisor. Finally, any displacement of the sheaf could be divided into two, three, or any number of parts, in the sense which I have given to this word above, and the divisor would still be part of the sheaf. The different displacements of the sheaf would be, so to speak, commensurable with one another. To every one of them would correspond a commensurable number, and vice versa. This therefore would be already a sort of mathematical continuity, but this continuity would still be imperfect, for there would be nothing corresponding to incommensurable numbers.

(3) We may suppose, finally, that our sheaf is perfectly continuous. All its displacements are interchangeable. To every commensurable or incommensurable number corresponds a displacement and vice versa. The displacement corresponding to the number na is nothing else than the displacement corresponding to the number a repeated n times. Why has the last of these three solutions been adopted? The reasons for the choice are complicated.

(1) It has been established by experience that displacements which are sufficiently large can be divided by any number whatever; and as the means of measurement increased in precision, this divisibility was demonstrated for displacements much smaller, with respect to which it first seemed doubtful. We have thus been led by induction to suppose that this divisibility is a property of all displacements, however small, and consequently to reject the first solution and to decide in favor of divisibility to infinity.

(2) The first solution, like the second, is incompatible with the other properties of the group which we know from other experience. I shall explain this further on. The third solution, accordingly, is imposed upon us by this fact alone. The contrary might have happened. It might have been that the properties of the group were incompatible with continuity. Then we should undoubtedly have adopted the first solution.

Sub-groups

The most important of the formal properties of a group is the existence of sub-groups. It must not be supposed that there can be as many sub-groups formed as we like, and that it is sufficient to cut up a group in an arbitrary manner, as one would inert clay, in order to obtain a sub-group. If two displacements be taken at random in a group, it will be necessary, in order to form a sub-group from them, to conjoin with them all their combinations and in the majority of cases it happens that in combining these two displacements in all possible manners we arrive ultimately at the primitive group again in its original intact form. It may happen thus that a group contains no sub-group.

But groups are distinguished from one another, in a formal point of view, by the number of sub-groups which they contain and by the mutual relations of the sub-groups. A superficial examination of the group of displacements readers it patent that it contains some sub-groups. A more minute examination will disclose them all. We shall see that among these sub-groups there are some that are: (1) continuous, i.e., have all their displacements divisible to infinity; (2) discontinuous, i.e., have no displacements that are divisible to infinity; (3) mixed, i.e., have displacements divisible to infinity and in addition others that are not so divisible.

From another point of view we distinguish among our sub-groups sheaves whose displacements are all interchangeable and those which do not possess this property.

The following is another manner of classing displacements and sub-groups.

Let us consider two displacements D and D'. Let D'' be a third displacement, defined to be the resultant of the displacement D' followed by the displacement D followed itself by the inverse displacement of D'. This displacement D'' is called the *transformation* of D by D'.[1]

From the formal point of view all the transformations of the same displacement are equivalent, so to speak; they play the same part; the Germans say that they are *geleichberechtigt.* Thus (if I may be permitted for an instant to use in advance the ordinary language of geometry which we are supposed not yet to know) two rotations of 60° are *geleichberechtigt*, two helicoidal displacements of the same step and same fraction of spiral are *geleichberechtigt.*[2]

The transformations of all displacements of a sub-group g by the same displacement D form a new sub-group which is called the transformation of the sub-group g by the displacement D. The different transformations of the same sub-group, playing the same part in a formal point of view, are *geleichberechtigt.*

It happens generally that many of the transformations of the same sub-group are identical; it will sometimes even happen that all the transformations of a sub-group are identical with one another and with the primitive sub-group. It is then said this sub-group is *invariant* (which happens, for example, in the case of the sub-group formed of all translations). The existence of an invariant sub-group is a formal property of the highest importance.[3]

Rotative sub-groups

The number of sub-groups is infinite; but they may be divided into a rather limited number of classes of which I do not wish to give here a complete enumeration. But these sub-groups are not all perceived with the same facility. Some among them have been only recently discovered. Their existence is not an intuitive truth. Unquestionably it can be (deduced from the fundamental properties of the group, from properties which are known to everybody, and which are, so to speak, the common patrimony of all minds. Unquestionably it is contained there in germ; yet those who have demonstrated their existence have justly felt that they had made a discovery and have frequently been obliged to write long memoirs to reach their results.

Other sub-groups, on the contrary, are known to us in much more immediate manner. Without much reflexion every one believes he has a direct intuition of them, and the affirmation of their existence constitutes the axioms of Euclid. Why is it that some sub-groups have directly attracted attention, whilst others have eluded all research for a much longer time? We shall explain it by a few examples.

A solid body having a fixed point is turning before our eyes. Its image is depicted on our retina and each of the fibres of the optic nerve conveys to us n impression; but owing to the motion of the solid body this impression is variable. One of these fibres, however, conveys to us a constant impression. It is that at the extremity of which the image of the fixed point has been formed. We have, thus, a change which causes certain sensations to vary, but leaves others invariable. This is a property of the displacement, but at first blush it does not appear that it is a formal property. It seems to belong to the qualitative character of the sensations experienced. We shall see, however, that we can disengage a formal property from it, and to render my thought clear I shall compare what takes place in this case with what happens in another instance which is apparently analogous.

I suppose that a certain body is moving before my eyes in any manner, but that a certain region of this body is painted in a color sufficiently uniform to leave no shades discernible. Let us say it is red. If the movements are not of too great compass and if the red region is sufficiently large in extent, certain parts of the retina will remain constantly in the image of that region, certain nerve-fibres will convey to us constantly the impression of the red, the displacement will have left certain sensations invariable.

But there is an essential difference between the two cases. Let us go back to the first one. We witnessed there an external change in which certain sensations A did not change, whilst other sensations B did change. We are able to correct this external change by an internal change, and in this correction the sensations A still remain invariable.

But now here is a new solid body which is turning before our eyes and is experiencing the same rotations as the first. This is a new external change which may be different altogether from the first from a qualitative point of view, because the new body which is turning may be painted in new colors, or because we are apprised of its rotation by touch and not by sight. We discover, however, that it is the *same* displacement, because it can be corrected by the same internal change. And we also discover that certain sensations A' in this new external change (totally different perhaps from A) have remained invariable, whilst other sensations B varied. Thus, this property of conserving certain sensations ultimately appears to us as a formal property independent of the qualitative character of these sensations.

We pass to the second example. We have, first, an external change in which a certain sensation C, a sensation of red, has remained constant. Let us suppose that another solid body, differently painted, undergoes the same displacement. Here is a new external change, and we know that it represents the same displacement because we can correct it by the same internal change. We discover generally that in this new external change certain sensations have not remained constant. Thus the conservation of the sensation C will appear to us as an accidental property only, connected with the qualitative nature of the sensation.

We are thus led to distinguish among displacements those which conserve certain sensations. The *ensemble* of the displacements which thus conserve a given system of sensations, evidently forms a sub-group which we may call a *rotative sub-group.*

Such is the conclusion which we draw from experience. It is needless to point out how crude is the experience and how precise on the other hand is the conclusion. Therefore experience cannot impose the conclusion upon us, but it suffices to suggest it to us. It suffices to show that of all the groups of which the models preexist in us, the only ones which we can accept with a view of referring to them our sensations, are those which contain such a sub-group. By the side of the rotative sub-group, we should consider its transformations, which also may he called rotative sub-groups. (Sub-group of rotations about a fixed point.) By new experiences, always very crude, it is then shown:

(1) That any two rotative sub-groups have common displacements.

(2) That these common displacements, all interchangeable among one another, form a sheaf, which may be called a rotative sheaf. (Rotations about a fixed axis.)

(3) That any rotative sheaf forms part not only of two rotative sub-groups, but of an infinity of them.

Here is the origin of the notion of the straight line, as the rotative sub-group was the origin of the notion of the point.

Let us now look at all the displacements of a rotative sheaf. If we look at any displacement whatever, it will not in general be interchangeable with all the displacements of the sheaf, but we shall discover very soon that there exist displacements which are interchangeable with all those of the rotative sheaf, and that they form a more extensive sub-group which may be called the helicoidal sub-group (combinations of rotations about an axis and of translations parallel to that axis). This will be evident when it is observed that a straight line can slide along itself.

Finally, we derive from the same crude observations such propositions as the following:

Any displacement sufficiently small and forming part of a given rotative sub-group, can always be decomposed into three others belonging respectively to three given rotative sheaves. Every displacement interchangeable with a rotative sub-group forms part of this sub-group.

Any displacement sufficiently small can always be decomposed into two others belonging respectively to two given rotative sub-groups, or to *six given rotative sheaves.*

Later on I shall revert in detail to the origin of these various propositions.

Translative sub-groups

With these propositions we have sufficient material, not to construct the geometry of Euclid, but to limit the choice between that of Euclid and the geometries of Lobachevsky and Riemann. In order to go farther, we are in need of a new proposition to take the place of the postulate of parallels. The proposition substituted will be the existence of an invariant sub-group, of which all the displacements are interchangeable and which is formed of all translations.

It is this that determines our choice in favor of the geometry of Euclid, because the group that corresponds to the geometry of Lobachevsky does not contain such an invariant sub-group.

Number of dimensions

In the ordinary theory of groups, we distinguish order and degree. Let us suppose the simplest case first, that of a group formed by different permutations between certain objects. The number of the objects is called the degree; the number of the permutations is called the order of the group. Two such groups may be isomorphic and their permutations may combine according to the same laws without their degree being the same. Thus let us consider the different ways in which a cube can be superposed upon itself. The vertices may be interchanged one with another, as may also be the faces and the edges; whence result three groups of permutations which are evidently isomorphic among themselves; but their degree may be either eight, six, or twelve, since there are eight vertices, six faces, and twelve edges.

On the other hand, two mutually isomorphic groups have always the same order. The degree is, so to speak, a material element, and the order a formal element, the importance of which is far greater. The theory of two groups of different degree may be the same so far as its formal properties are concerned; just as the mathematical theory of the addition of three cows and four cows is identical with that of three horses and four horses.

When we pass to continuous groups, the definitions of order and degree must be modified, though without sacrificing their spirit. Mathematicians suppose ordinarily that the object of the operations of the group is an *ensemble* of a certain number n of quantities susceptible of being varied in a continuous manner, which quantities are called *co-ordinates.* On the other hand, every operation of the group may be regarded as forming part of a sheaf analogous to the rotative sheaf and as a multiple of a very high order of an infinitesimal operation belonging to the same sheaf. Then, every infinitesimal operation of the group can be decomposed into k other operations belonging to k given sheaves. The number n of the co-ordinates (or of the dimensions) is then the degree, and the number k of the components of an infinitesimal operation is the *order.* Here again two isomorphic groups may have different degrees, but must be of the same order. Here again the degree is an element relatively material and secondary, and the order a formal element. According to the laws established above, our group of displacements is here of the sixth order, but its degree is yet unknown. Is the degree given us immediately?[7]

Displacements, we have seen, correspond to changes in our sensations, and if we distinguish in the present group between form and material, the material can be nothing else than that which the displacements cause to change, viz., our sensations. Even if we suppose that what we have above called sensible space has already been elaborated, the material would then be represented by as many continuous variables as there are nerve-fibres; the "degree" of our group would then be extremely large. Space would not have three dimensions but as many as there are nerve-fibres. Such is the consequence to which we come if we accept as the material of our group what is immediately given us. How shall we escape the difficulty? Evidently by replacing the group which is given us, together with its form and its material, by another *isomorphic* group, the material of which is simpler.

But how is this to be done? It is precisely owing to this circumstance, that the displacements which conserve certain elements are the same as those which conserve certain other elements. Then all those elements which are conserved by the same displacements we agree to replace by a single element which has a purely schematic value only. Whence results a considerable reduction of degree.

For example, I see a solid body rotating about a fixed point. The parts near the fixed point are painted red. Here is a displacement, and within this displacement I perceive that something remains invariable—namely, the sensation of red conveyed to me by a certain optical nerve-fibre. Some time afterward I see another solid body turning about a fixed point. But the parts near the fixed point are painted green. The sensations experienced are in themselves quite different, but I perceive that it is the same displacement because it can be corrected by the same internal change. Here again something remains invariable; but this something is totally different from the material point of view; it is the sensation of green conveyed by a certain nerve-fibre.

These two things, which materially are so different, I replace schematically by a single thing which I call a point, and I express my thought by saying that in the one case as in the other, a point of the body has remained fixed. Thus every one of our new elements will be what is conserved by all the displacements of a sub-group; to every sub-group there will then correspond an element and vice versa.

Let us consider the different transformations of the same sub-group. They are infinite in number and may form a simple, double, triple, continuous infinity. To each one of these transformations an element can be made to correspond; I have then a simple, double, triple, etc., infinity of them, and the degree of our continuous group is $1, 2, 3, \ldots$.

Suppose that we choose the different transformations of a rotative sub-group. We have here a triple infinity. The material of our group is accordingly composed of a triple infinity of elements. The degree of the group is three. We have then chosen the point as the element of space and given to space three dimensions.

Suppose we choose the different transformations of a helicoidal sub-group. Here we have a quadruple infinity. The material of our group is composed of a quadruple infinity of elements. Its degree is four. We then have chosen the straight line as the element of space,—which would give to space four dimensions.

Suppose, finally, that we choose the different transformations of a rotative sheaf. The degree would then be five. We have chosen as the element of space the figure formed by a straight line and a point on that straight line. Space would have five dimensions.

Here are three solutions, which are each logically possible. We prefer the first because it is the simplest, and it is the simplest because it is that which gives to space the smallest number of dimensions. But there is another reason which recommends this choice. The rotative sub-group first attracts our attention because it conserves certain sensations. The helicoidal sub-group is known to us only later and more indirectly. The rotative sheaf on the other hand is itself merely a sub-group of the rotative sub-group.

The notion of point

I feel that I am here touching on the most delicate spot of this discussion, and I am compelled to stop for a moment to justify more completely my previous assertions which same persons may be disposed to doubt. Many persons, indeed, would consider the notion of a point of space as so immediate and so clear that any definition of it is superflous. But I believe it will be granted me that so subtle a notion as that of the mathematical point, without length, breadth, or thickness, is not immediate, and that it needs to be explained.

But is it the same with the vaguer and less precisely defined, yet more empirical notion, of *place*? Is there anyone who does not fancy he knows perfectly well what he is talking about when he says: this object occupies the place which was just occupied by that object. To determine the range of such an assertion, and the conclusions which can be drawn from it, let us seek to analyse its signification. If I have moved neither my body, my head, nor my eye, and if the image of the object B affects the same retinal fibres that the image of the object A previously affected; if again, although I have moved neither my arm nor my hand, the same sensory fibres which extend to the end of the finger, and which formerly conveyed to me the impression which I attributed to the object A now convey to me the impression which I attribute to the object B; if both these conditions are fulfilled,—then ordinarily we agree to say that the object B occupies the place which previously the object A occupied.

Before analysing so complicated a convention as that just stated I shall first make a remark. I have just enunciated two conditions: one relating to sight, and one relating to touch. The first is necessary but not sufficient, for we say in ordinary language that the point on the retina where an image is formed gives us knowledge only of the direction of the visual ray, but that the distance from the eye remains unknown. The second condition is at once necessary and sufficient, because we assume that the action of touch is not exercised at a distance, and that the object A like the object B cannot act upon the finger except by immediate contact. All this agrees with what experience has taught us; namely, that the first condition can be fulfilled without the second being realised, but that the second cannot be fulfilled without the first. Let it be remarked that we have here something which we could not know a priori, that experience alone is able to demonstrate it to us.

Nor is this all. To determine the place of an object I made use only of an eye and a finger. I could have made use of several other means,—for example, of all my other fingers. Having been made aware that the object A has produced upon my first finger a tactual impression, suppose that by a series of movements S my second finger comes into contact with the same object A. My first tactual impression ceases and is replaced by another tactual impression which is conveyed to me by the nerve of the second finger, and which I still attribute to the action of the object A. Some time afterwards, and without my having moved my hand, the same nerve of the second finger conveys to me another tactual impression, which I attribute to the action of another object B. I then say that the object B has taken the place of the object A.

At this moment I make a series of movements S' the inverse of the series S. How do I know that these two series are inverse to one another? Because experience has taught me that when the internal change S that corresponds to certain muscular sensations is followed by an internal change S' which corresponds to other muscular sensations, a compensation is effected and my primitive impressions, originally modified by the change S, are reestablished by the change S'.

I execute the series of movements S'. The effect ought to be to take back my first finger to its initial position and so to put it into contact with the object B, which has taken the place of the object A. I ought, therefore, to expect that the nerve of my first finger should convey to me a tactual sensation attributable to the object B. In fact this is what happens.

But would it therefore be absurd to suppose the contrary? And why would it be absurd? Shall I say that the object B having taken the place of the object A, and my first finger having resumed its original place, it ought to touch the object B just as before it touched the object A? This would be an outright begging of the question. And to show this let us attempt to apply the same reasoning to another example, or rather let us return to the example of sight and touch which I cited at the outset.

The image of the object A has made an impression on one of my retinal fibres. At the same time the nerve of one of my fingers conveys to me a tactual impression which I attribute to the same object. I move neither my eye nor my hand. And a moment after the image of the object B has impressed the same retinal fibre. By a course of reasoning perfectly similar to that which precedes, I should be tempted to conclude that the object B had taken the place of the object A, and I should expect that the nerve of my finger would convey to me a tactual impression attributable to B. And yet I should be deceived. For the image of B may chance to be formed upon the same point of the retina as the image of A, although the distance to the eye may not be the same in the two cases.

Experience has refuted my reasoning. I extricate myself by saying that it is not sufficient for two bodies to cast their image upon the same retinal fibre in order to justify me in saying that the two bodies are in the same place; and I should extricate myself in a similar manner in the case of the two fingers, if the indications of the second finger had not been in accord with those of the first, and if experience had been at variance with my reasoning. I should still say that two objects A and B can make an impression upon the same finger by means of touch and yet not be in the same place; in other words, I should conclude that touch could be effected at a distance. Or, again, I should agree to consider A and B as being in the same place only on the condition of there being concordance not only between their effects upon the first finger, but also between their effects upon the second finger. One might almost say, in a certain point of view, that one more dimension would be attributed to space in this manner.

To sum up, there are certain laws of concordance, which can be revealed to us only by experience, and which are at the basis of the vague notion of place.

But even taking these laws of concordance for granted, can we deduce from them the much more precise notion of point and the notion of number of dimensions? This remains to be examined.

First an observation. We have spoken of two objects A and B, which have cast one after another their image on the same point of the retina. But these two images are not identical; otherwise how could I distinguish them? They differ, for example, in color. The one is red, the other is green. We have, accordingly, two sensations which differ in quality and which are doubtless conveyed to me by two different though contiguous nerve-fibres. What have they in common with one another, and why am I led to associate them together? I believe that if the eye were immovable, we should never have thought of this association. It is the movements of the eye that have taught us that there is the same relation on the one hand between the sensation of green at the point A of the retina and the sensation of green at the point B of the retina, and on the other hand between the sensation of red at the point A of the retina and the sensation of red at the point B of the retina. We have found, in fact, that the same movements, corresponding to the same muscular sensations, cause us to pass from the first to the second, or from the, third to the fourth. Were this not so, these four sensations would appear qualitatively distinct, and we should no more think of establishing a sort of proportion between them than we should between an olfactory, a gustatory, an auditive and a tactual sensation.

Yet whatever be the origin of this association, it is implied in the notion of place, which could not have grown up without it. Let us analyse, therefore, its laws. We can only conceive them under two different forms equally remote from mathematical continuity; namely, under the form of discontinuity or under the form of physical continuity.

Under the first form, our sensations will be divided into a very large number of "families"; all the sensations of one family being associated with one another and not being associated with those of other families. Since to every family there would correspond a place, we should have a finite but very large number of places, and the places would form a discrete aggregate. There would be no reason for classifying them in a table of three dimensions rather than in one of two or four; and we could not deduce from them either the mathematical point or space.

Under the second form, which is more satisfactory, the different families interpenetrate one another. A, for example, will be associated with B, and B with C. But A will not appear to us as associated with C. We shall find that A and C do not belong to the same family, although on the one hand A and B, and on the other hand B and C, will appear to us as belonging to the same family. Thus we cannot distinguish between a weight of nine grammes and one of ten grammes, or between the latter weight and a weight of eleven grammes. But we can readily tell the difference between the first weight and the third. This is always the formula of physical continuity.

Let us picture to ourselves a series of wafers partially covering one another in such wise that the plane is totally covered; or better, let us picture to ourselves something analogous in a space of three dimensions. If these wafers were to form by their superposition only a sort of one-dimensional ribbon, we should recognize it, because the associations of which I have just been speaking obey a law that may be stated as follows: if A is associated at once with B, C, and D, D is associated with B or with C. This law would not be true if our wafers covered by their superposition a plane or a space of more than two dimensions. When I say, therefore, that all possible places constitute an aggregate of one dimension or of more than one dimension, I mean to say simply that this law is true or that it is false. When I say that they constitute an aggregate of two or three dimensions, I simply affirm that certain analogous laws are true.

Such are the foundations on which we may attempt to construct a *static* theory of the number of dimensions. It will be seen how complicated is this manner of defining the number of dimensions, how imperfect it is, and it is useless to remark upon the distance which still separates the physical continuity of three dimensions as thus understood from the real mathematical continuity of three dimensions.

Discussion of the preceding theory

Without dwelling upon the multitude of difficult details, let us see in what those associations consist upon which the notion of place rests. We shall see that we are finally led back, after a long detour, to the notion of group, which appeared to us at the outset the best fitted for elucidating the question of the number of dimensions.

By what means are different "places" distinguished from one another? How, for example, are two places occupied successively by the extremity of one of my fingers to be distinguished? Evidently by the movement which my body has made in the interval, movements which are made known to me by a certain series of muscular sensations. These two places correspond to two distinct attitudes and positions of the body which are known solely by the movements which I have had to make in changing a certain initial attitude and a certain initial position; and these movements themselves are known to me only by the muscular sensations which they have provoked.

Two attitudes of the body, or two corresponding places of the finger, appear to me identical if the two movements which I must make to reach them differ so little from each other that I cannot distinguish the corresponding muscular sensations. They will appear to me non-identical, without some new convention, if they correspond to two series of distinguishable muscular sensations.

But in this manner we have engendered not a physical continuity of three dimensions but a physical continuity of a much larger number of dimensions; for I can cause the muscular sensations corresponding to a very large number of muscles to vary, and I do not on the other hand consider a single muscular sensation only, nor even an aggregate of simultaneous sensations, but a series of successive sensations, and I can make the laws by which these sensations succeed one another vary in an arbitrary manner.

Why is the number of dimensions reduced, or, what is the same thing, why do we consider two places as identical when the two corresponding attitudes of the body are different? Why do we say in certain cases that the place occupied by the extremity of a finger has not changed, although the attitude of the body has changed?

It is because we discover that very *frequently*, in the movement which causes the passage from the one to the other of these two attitudes, the tactual sensation attributable to the contact of this finger with an object A persists and remains constant. We *agree* then, to say that these two attitudes shall be placed in the same class and that this class shall embrace all attitudes corresponding to the same place occupied by the same finger. We agree that these two attitudes shall still be placed in the same class even when they are accompanied by no tactual sensation, or by variable tactual sensations.

This convention has been evoked by experience, because experience alone informs us that certain tactual sensations are frequently persistent. But in order that conventions of this kind shall be permissible, they must satisfy certain conditions which it now remains for us to analyse.

If I place the attitudes A and B in the same class, and also the attitudes B and C in the same class, it follows necessarily that the attitudes A and C must be regarded as belonging to the same class. If, then, we agree to say that the movements which cause the passage from the attitude A to the attitude B do not change the place of the finger, and if the same holds true of the movements which cause the passage from the attitude B to the attitude C, it follows necessarily that the same must again be true of those which cause the passage from the attitude A to the attitude C. In other words, the aggregate of the movements causing a passage from one attitude to another attitude of the same class constitutes a group. It is only when such a group exists that the convention above laid down is acceptable. To every class of attitudes, and consequently to every place, there will therefore correspond a group, and we are here led back again to the notion of group, without which there would be no geometry.

Nevertheless, there is a difference between the principle here under discussion and the theory which I developed above. Here each place appears to me associated with a certain group which is introduced as the sub-group S of the group G formed by the movements which can give to the body all possible positions and all possible attitudes, the relative situations of the different parts of the body being allowed to vary in any manner whatsoever. In our other theory, on the contrary, every point was associated with a sub-group S' of the group G' formed by the displacements of the body viewed as an invariable solid, that is to say, by displacements such that the relative situations of the different parts of the body do not vary.

Which of the two theories is to be preferred? It is evident that G' is a sub-group of G and S' a sub-group of S. Further, G' is much simpler than G, and for this reason the theory which I first propounded and which is based upon the consideration of the group G' appears to me simpler and more natural, and consequently I shall hold to it.

But be this as it may, the introduction of a group, more or less complicated, appears to be absolutely necessary. Every purely statical theory of the number of dimensions will give rise to many difficulties, and it will always be necessary to fall back upon a dynamical theory. I am happy to be in accord on this point with the ideas set forth by Professor Newcomb in his *Philosophy of Hyperspace* [Newcomb 1898].

The reasoning of Euclid

But in order to show that the idea of displacement, and consequently the idea of group, has played a preponderant part in the genesis of geometry, it remains to be shown that this idea dominates all the reasoning of Euclid and of the authors who after him have written upon elementary geometry.

Euclid begins by enunciating a certain number of axioms; but it must not be imagined that the axioms which he enunciates explicitly are the only ones to which he appeals. If we carefully analyse his demonstrations we shall find in them, in a more or less masked form, a certain number of hypotheses which are in reality axioms disguised; and we may say almost as much of some of his definitions.

His geometry begins with declaring that two figures arc equal if they are superposable. This assumes that they can be displaced and also that among all the changes which the may undergo, we can distinguish those which may be regarded as displacements without deformation. Again, this notion implies that two figures which are equal to a third are equal to each other. And that is tantamount to saying that if there be a displacement which puts the figure A upon the figure B, and a second displacement which superposes the figure B upon the figure C, there will also be a third, the resultant of the first two, which will superpose the figure A upon the figure C. In other words, it is presupposed that the displacements form a group. The notion of a group, accordingly, is introduced from the outset, and inevitably introduced.

When I pronounce the word "length," a word which we frequently do not think necessary to define, I implicitly assume that the figure formed by two points is not always superposable upon that which is formed by two other points; for otherwise any two lengths whatever would be equal to each other. Now this is an important property of our group.

I implicitly enunciate a similar hypothesis when I pronounce the word "angle."

And how do we proceed in our reasonings? By displacing our figures and causing them to execute certain movements. I wish to show that at a given point in a straight line a perpendicular can always be erected, and to accomplish this I conceive a movable straight line turning about the point in question. But I presuppose here that the movement of this straight line is possible, that it is continuous, and that in so turning it can pass from the position in which it is lying on the given straight line, to the opposite position in which it is lying on its prolongation. Here again is a hypothesis touching the properties of the group.

To demonstrate the cases of the equality of triangles, the figures are displaced so as to be superposed one upon the other.

Finally, what is the method employed in demonstrating that from a given point one and only one perpendicular can always be drawn to a given straight line? The figure is turned 180° around the given straight line, and in this manner the point symmetrical to the given point with respect to the given straight line is obtained. We have here a feature most characteristic, and here appears the part which the straight line most frequently plays in geometrical demonstrations, namely, that of an axis of rotation.

There is implied here the existence of the sub-group which I have called the rotative sheaf. When—which also frequently happens—a straight line is made to slide along itself (for we shall, of course, continue to suppose that it can serve as an axis of rotation), we implicitly take the existence of the helicoidal sub-group for granted. In fine, the principal foundation of Euclids demonstrations is really the existence of the group and its properties.

Unquestionably he appeals to other axioms which it is more difficult to refer to the notion of group. An axiom of this kind is that which some geometers employ when they define a straight line as the shortest distance between two points. *But it is precisely such axioms that Euclid enunciates.* The others, which are more directly associated with the idea of displacement and with the idea of groups, are the very ones which he implicitly admits, and which he does not deem it even necessary to enunciate. This is tantamount to saying that the former are the fruit of a later experience, that the others were first assimilated by us, and that consequently the notion of group existed prior to all the others.

The geometry of Staudt

It is known that Staudt attempted to base geometry upon different principles. Staudt admits the following axioms only:

1. Through two points a straight line can always be drawn.
2. Through three points a plane can always be drawn.
3. Every straight line which has two of its points in a plane lies entirely in that plane.
4. If three planes have one point in common, and one only, any straight line will cut at least one of these three planes.

These axioms are sufficient to establish all the *descriptive* properties relating to the intersections of straight lines and planes. To obtain the metrical properties we begin with *defining* a harmonic pencil of four straight lines, taking as definition the well-known descriptive property. Then the anharmonic ratio of four points is *defined*, and finally, supposing that one of these four points has been relegated to infinity, the ratio of two lengths is *defined*.[4]

This last is the weak point of the foregoing theory, attractive though it be. To arrive at the notion of length by regarding it merely as a particular case of the anharmonic ratio is an artificial and repugnant detour. This evidently is not the manner in which our geometrical notions were formed.

Let us see now whether we can conceive, without the introduction of the notion of group and of movement, how the notions which serve as the foundations of this ingenious geometry have taken their rise. Let us see what experiences might have led us to formulate the axioms enunciated above.

If the straight line is not given as an axis of rotation, it can be given only in one way, namely, as the trajectory of a ray of light. I mean, that the experiences, always more or less crude, which serve us as our point of departure, should all be applicable to the ray of light, and that we must define the straight line as a line for which the simple laws which the ray of light approximately obeys will be rigorously true. The following is the experience which must be made in order to verify the most important of our axioms, namely, the third.

Let two threads be stretched. Let the eye be placed at the extremity of one of these threads. We see that the thread is entirely hidden by its extremity, which teaches us that the thread is rectilinear, that is to say, is the direction of the trajectory of a ray of light. Let the same be done for the second thread. The following is then observed: either there will be no position of the eye for which one of the threads is entirely hidden by the other, or there will he an infinity of such positions.

How is the question of the number of dimensions presented in this order of ideas? Let us consider all the positions of the eye for which one of the strings is hidden by the other. Let us suppose that in one of these positions the point A of the first string is hidden by the point A' of the second, the point B by the point B', the point C by the point C'. We then discover that if the body is so displaced that the point A is always hidden by the point A' and the point B by the point B', that the point C always remains hidden by the point C', and that in general any point whatsoever of the first thread remains hidden by the same point of the second thread by which it was hidden before the body was displaced. We express this fact by saying that although the body is displaced, the position of the eye has not changed.

We see thus that the position of the eye is defined by two conditions, viz., that A is hidden by A' and B by B'. We express this fact by saying that the locus of the points such that the two threads mutually hide each other has two dimensions.

Similarly, let us suppose that in a certain position of the body four threads A, B, C, D, hide four points A', B', C', D'; let us suppose that the body is displaced, but in such a manner that A, B, and C continue to hide A', B', and C'. We shall then discover that D continues to hide D', and we shall again express this fact by saying that the position of the eye has not changed. This position will therefore be defined by three conditions, and this is why we say that space has three dimensions.

It will be remarked that the law as thus experimentally discovered, is only approximately true. But this is not all. It is not even always true, because D or D' may have moved at the same time that my body was being displaced. We then simply declare that this law is often approximately true.

But we are desirous of arriving at geometrical axioms which are rigorously and always true, and we always escape the dilemma by the same artifice, namely, by saying that we agree to consider the change observed as the resultant of two others, viz., of one which rigorously obeys the law and which we attribute to the displacement of the eye, and of a second one which is generally very small and which we attribute either to qualitative alterations or to the movements of external bodies.

We have not been able to avoid the consideration of movements of the eye and of the body, yet we may say, that from a certain point of view the geometry of Staudt is predominantly a visual geometry, while that of Euclid is predominantly muscular.

Undoubtedly unconscious experiences analogous to those of which I have just spoken may have played a part in the genesis of geometry; but they are not sufficient. If we had proceeded, as the geometry of Staudt supposes us to have done, some Apollonius would have discovered the properties of polars.[5] But it would have been only long after, that the progress of science would have made clear what a length or an angle is. We should have had to wait for some Newton to discover the various cases of the equality of triangles. And this is evidently not the way that things have come to pass.

The axiom of Lie

It is Sophus Lie who has contributed most towards making prominent the importance of the notion of group and laying the foundations of the theory that I have just expounded. It is he, in fact, who gave the present form to the mathematical theory of continuous groups. But to render possible its application to geometry, he regards a new axiom as necessary, which he enunciates by declaring that space is a *Zahlenmannigfaltigkeit*, that is, that to every point of a straight line there corresponds a number and vice versa.[6]

Is this axiom absolutely necessary? And could not the other principles which Lie has laid down dispense with it? We have seen above in connexion with continuity, that the best known groups may be distributed from a certain point of view into three classes; all the operations of the group can be divided into sheaves; for "discontinuous" groups the different operations of the same sheaf are only a single operation repeated once, twice, three times, etc.; for "continuous" groups properly so called the different operations of the same sheaf correspond to different whole numbers, commensurable or incommensurable; finally, for groups which may be called "semi-continuous," these operations correspond to different commensurable numbers.

Now it may be demonstrated that no discontinuous or semi-continuous group exists possessing other properties than those which experience has led us to adopt for the fundamental group of geometry, and which I here briefly recall: The group contains an infinity of sub-groups, all *gelichberechtigt*, which I call rotative sub-groups. Two rotative sub-groups have a sheaf in common which I call rotative and which is common not only to two but also to an infinity of rotative sub-groups. Finally, every very small displacement of the group may be regarded as the resultant of six displacements belonging to six given rotative sheaves. A group satisfying these conditions can be neither discontinuous nor semi-continuous.[7]

Unquestionably this is an exceedingly recondite property, and not easy to demonstrate. Geometers who were ignorant of it have not the less hit upon its consequences, as for example, when they learned that the ratio of a diagonal to the side of a square is incommensurable. It was for this reason that the introduction of incommensurables into geometry became necessary.

The group, therefore, must be continuous, and it seems as if the axiom of Lie were useless.

Nevertheless, we are obliged to remark that the classification of groups above sketched is not complete; groups may be conceived which are not included in it. We might, therefore, suppose that the group is neither discontinuous, semi-continuous, nor continuous. But this would be a complex hypothesis. We reject it, or rather we never think of it, for the reason that it is not the simplest compatible with the axioms adopted.

The foundation of the axiom of Lie remains to be supplied.

Geometry and contradiction

In following up all the consequences of the different geometrical axioms, are we never led to contradictions? The axioms are not analytical judgments a priori; they are conventions. Is it certain that all these conventions are compatible?

These conventions, it is true, have all been suggested to us by experiences, but by crude experiences. We discover that certain laws are approximately verified, and we decompose the observed phenomenon conventionally into two others: a purely geometrical phenomenon which exactly obeys these laws; and a very minute disturbing phenomenon.

Is it certain that this decomposition is always permissible? It is certain that these laws are approximately compatible, for experience shows that they are all approximately realised at one and the same time in nature. But is it certain that they would be compatible if they were absolutely rigorous?

For us the question is no longer doubtful. Analytical geometry has been securely established, and *all* the axioms have been introduced into the equations which serve as its point of departure; we could not have written these equations if the axioms had been contradictory. Now that the equations are written, they can be combined in all possible manners; analysis is the guarantee that contradictions shall not be introduced.

But Euclid did not know analytical geometry, and yet he never doubted for a moment that his axioms were compatible. Whence came his confidence? Was he the dupe of an illusion? And did he attribute to our unconscious experiences more value than they really possess? Or perhaps, since the idea of the group was potentially pre-existent in him, did he have some obscure instinct for it, without reaching a distinct notion of it? I shall leave the question undecided although inclined towards the second solution.

The use of figures

It may be asked why geometry cannot be studied without figures. This is easy to account for. When we commence studying geometry, we have already had in innumerable instances the fundamental experiences which have enabled our notion of space to originate. But they were made without method, without scientific attention and unconsciously, so to speak. We have acquired the ability *to represent to ourselves* familiar geometrical experiences without being obliged to resort to material reproductions of them; but we have not yet deduced from them logical conclusions. How is this to be done? Before enunciating the law, the experience in question is perceptually represented by stripping it as completely as possible of all accessory or disturbing circumstances,—just as a physicist eliminates the sources of systematic error in his experiments. It is here that figures are necessary, but they are an instrument only slightly less crude than the chalk which is employed in drawing them; and, like material objects, it is beyond our power to represent them in the geometrical space which forms the object of our studies; we can only represent them in sensible space. We accordingly do not study material figures, but simply make use of them in studying something which is higher and more subtle.

Form and matter

We owe the theory which I have just sketched to Helmholtz and Lie. I differ from them in one point only, but probably the difference is in the mode of expression only and at bottom we are completely in accord.

As I explained above, we must distinguish in a group the form and the matter [material]. For Helmholtz and Lie the matter of the group existed previously to the form, and in geometry the matter is a *Zahlenmannigfaltigkeit* of three dimensions. *The number of dimensions is therefore posited prior to the group.* For me, on the contrary, the form exists before the matter. The different ways in which a cube can be superposed upon itself, and the different ways in which the roots of a certain equation may be interchanged, constitute two isomorphic groups. They differ in matter only. The mathematician should regard this difference as superficial, and he should no more distinguish between these two groups than he should between a cube of glass and a cube of metal. In this view the group exists prior to the number of dimensions.

We escape in this way also an objection which has often been made to Helmholtz and Lie. "But your group," say these critics, "presupposes space; to construct it you are obliged to assume a continuum of three dimensions. You proceed as if you already knew analytical geometry." Perhaps the objection was not altogether just; the continuum of three dimensions which Helmholtz and Lie posited was a sort of non-measurable magnitude analogous to magnitudes concerning which we may say that they have grown larger or smaller, but not that they have become twice or three times as large.

It is only by the introduction of the group, that they made of it a measurable magnitude, that is to say a veritable space. Again, the origin of this non-measurable continuum of three dimensions remains imperfectly explained.

But, it will be said, in order to study a group even in its formal properties, it is necessary to construct it, and it cannot be constructed without matter. One might as well say that one cannot study the geometrical properties of a cube without supposing this cube to be of wood or of iron. The complexus of our sensations has without doubt furnished us with a sort of matter, but there is a striking contrast between the grossness of this matter and the subtle precision of the form of our group. It is impossible that this can be, properly speaking, the matter of such a group. The group of displacements such as it is given us directly by experience, is something more gross in character; it is, we may say, to continuous groups proper what the physical continuum is to the mathematical continuum. We first study its form agreeably to the formula of the physical continuum, and since there is something repugnant to our reason in this formula we reject it and substitute for it that of the continuous group which, potentially, pre-exists in us, but which we originally know only by its form. The gross matter which is furnished us by our sensations was but a crutch for our infirmity, and served only to force us to fix our attention upon the pure idea which we bore about in ourselves previously.

Conclusions

Geometry is not an experimental science; experience forms merely the occasion for our reflecting upon the geometrical ideas which pre-exist in us. But the occasion is necessary; if it did not exist we should not reflect; and if our experiences were different, doubtless our reflexions would also be different. Space is not a form of our sensibility; it is an instrument which serves us not to represent things to ourselves, but to reason upon things.

What we call geometry is nothing but the study of formal properties of a certain continuous group; so that we may say, space is a group. The notion of this continuous group exists in our mind prior to all experience; but the assertion is no less true of the notion of many other continuous groups; for example, that which corresponds to the geometry of Lobachevsky. There are, accordingly, several geometries possible, and it remains to be seen how a choice is made between them. Among the continuous mathematical groups which our mind can construct, we choose that which deviates least from that rough group, analogous to the physical continuum, which experience has brought to our knowledge as the group of displacements.

Our choice is therefore not imposed by experience. It is simply guided by experience. But it remains free; we choose this geometry rather than that geometry, not because it is more *true*, but because it is the more *convenient*.

To ask whether the geometry of Euclid is true and that of Lobachevsky is false, is as absurd as to ask whether the metric system is true and that of the yard, foot, and inch, is false. Transported to another world we might undoubtedly have a different geometry, not because our geometry would have ceased to be true, but because it would have become less convenient than another. Have we the right to say that the choice between geometries is imposed by reason, and, for example, that the Euclidean geometry is alone true because the principle of the relativity of magnitudes is inevitably imposed upon our mind? It is absurd, they say, to suppose a length can be equal to an abstract number. But why? Why is it absurd for a length and not absurd for an angle? There is but one answer possible. It appears to us absurd, because it is contrary to our habitual way of thinking. Unquestionably reason has its preferences, but these preferences have not this imperative character. It has its preferences for the simplest because, all other things being equal, the simplest is the most convenient. Thus our experiences would be equally compatible with the geometry of Euclid and with a geometry of Lobachevsky which supposed the curvature of space to be very small. We choose the geometry of Euclid because it is the simplest. If our experiences should be considerably different, the geometry of Euclid would no longer suffice to represent them conveniently, and we should choose a different geometry.

Let it not be said that the reason why we deem the group of Euclid the simplest is because it conforms best to some pre-existing ideal which has already a geometrical character; it is simpler because certain of its displacements are interchangeable with one another, which is not true of the corresponding displacements of the group of Lobachevsky. Translated into analytical language, this means that there are fewer terms in the equations, and it is clear that an algebraist who did not know what space or a straight line was would nevertheless look upon this as a condition of simplicity.

In fine, it is our mind that furnishes a category for nature. But this category is not a bed of Procrustes into which we violently force nature, mutilating her as our needs require. We offer to nature a choice of beds among which we choose the couch best suited to her stature.

Notes

[Reprinted without change from Poincaré 1898, which lists the translator as T. J. McCormack. For a detailed account and critique of this paper, see Torretti 1978, 340–352, who argues that "Poincaré's entire construction rests upon an untenable theory of perception, according to which all our knowledge of physical facts can be ultimately traced to a variegated and changing aggregate of elementary sensations, each of whcih is caused by the momentary stimulation of an afferent nerve. Because they rest on such foundations, many statements in the essay are unclear or simply unlikely ..." (340–341).]

1. [Here Poincaré is constructing what presently is called an *equivalence class*, for which he also uses the German term *gleichberechtigt* (literally, "having equal right"). Symbolically, we can write what he is saying as defining $D'' = D' * D * D'^{-1}$, where the operation of the group is called $*$ and D'^{-1} is the "inverse" of the element D', so that (by definition) $D' * D'^{-1} = I$, the identity element contained in all groups (so that $D * I = I * D = D$ for any D). In words, what his "transformation" means is that D'' takes a certain D and first applies to it the element D' and then "undoes" that by applying the inverse of that same element, D'^{-1}. The idea behind this procedure is to break up the group into classes (now called "cosets") of equivalent elements. For some more details, see Pesic 2003, 176–177; for more detailed discussion, see Maxfield and Maxfield 1992, 26–29, 62–69.]

2. [By "helicoidal," Poincaré means motion in a helical spiral, a rotating circle that also advances along an axis perpendicular to its center, sweeping out a cylindrical path, as with the advancing motion of screw, for instance.

3. [These invariant or "normal" sub-groups are important because they include (roughly speaking) all of a certain similar class of group elements, similar in the sense of being equivalent. For instance, in the case of rotations of a equilateral triangle by 60°, an invariant sub-group would need to include all possible such rotations, not just one of them. See Maxfield and Maxfield 1992, 70–77, and Pesic 2003, 193–195.]

4. [For the definition of the anharmonic ratio, see above 114, note 1.]

5. [Apollonius of Perga (third century B. C.) wrote an important work on conic sections.]

6. [The term *Zahlenmannigfaltigkeit* literally means "number manifold"; here, Poincaré uses it to mean the identification of each point on a line with a real number.]

7. [For the number six characterizing this rotative group, see above, 115, note 3.]

Geometry and Experience (1921)

Albert Einstein

Mathematics enjoys special esteem above all other sciences for *one* reason: its propositions are absolutely certain and indisputable, while those of all other sciences are to some extent debatable and in constant danger of being overthrown by newly discovered facts. Despite this, the researcher in another domain of science would not need to envy the mathematician if the propositions of mathematics referred not to objects of reality but to objects purely of our imagination. For it is no wonder that different persons should arrive at the same logical conclusions when they have already agreed upon the fundamental propositions (axioms), as well as the methods by which other propositions are to be deduced. But there is another reason for the high repute of mathematics, in that it is mathematics which affords the exact natural sciences a certain measure of security, to which without mathematics they could not attain.

At this point a riddle presents itself that has troubled researchers throughout the ages. How is it possible that mathematics, being after all a product of human thought that is independent of experience, is so admirably appropriate to the objects of reality? Can human reason, then, without experience, through pure thought fathom the properties of real things?

In my opinion, the answer to this question is, briefly, this: As far as the laws of mathematics refer to reality, they are not certain, and as far as they are certain, they do not refer to reality. It seems to me that complete clarity as to this state of things first became common property through that trend in mathematics known by the name of "axiomatics."[1] The progress achieved by axiomatics consists in its having neatly separated the logical-formal from its objective or intuitive content; according to axiomatics, the logical-formal alone forms the subject matter of mathematics, which is not concerned with the intuitive or other content associated with the logical-formal.

Let us for a moment consider from this point of view any axiom of geometry, for instance, the following: Through two points in space there passes one and only one straight line. How is this axiom to be interpreted in the older and in the newer sense?

The older interpretation: Every one knows what a straight line is and what a point is. Whether this knowledge springs from an ability of the human mind or from experience, from some collaboration of the two or from some other source, is not for the mathematician to decide. He leaves the question to the philosopher. Being based upon this knowledge, which precedes all mathematics, the axiom stated above is (like all other axioms) self-evident, that is, it is the expression of a part of this a priori knowledge.

The newer interpretation: Geometry treats of objects denoted by the words straight line, point, etc. No knowledge or intuition whatever of these objects is assumed, but only the validity of the axioms, such as the one stated above, which are to be taken in a purely formal sense, that is, as void of all content of intuition or experience. These axioms are free creations of the human mind. All other propositions of geometry are logical inferences from the axioms (which are to be taken in the nominalistic sense only). The axioms first define the objects of which geometry treats. In his book on epistemology, Schlick has therefore characterized axioms very aptly as "implicit definitions."[2]

This view of axioms, advocated by modern axiomatics, purges mathematics of all extraneous elements, and thus dispels the mystical obscurity that formerly surrounded the foundations of mathematics.[3] But such a presentation of its foundations thus clarified also makes it evident that mathematics as such cannot predicate anything about objects of our intuition or real objects. In axiomatic geometry, the words "point," "straight line," etc. are understood only as empty conceptual schemata. That which gives them content does not belong to mathematics.

Yet on the other hand it is certain that mathematics generally, and particularly geometry, owes its existence to the desire to learn something about the behavior of real things. The very word "geometry," which, of course, means earth-measuring, proves this. For earth-measuring has to do with the possibilities of the relative arrangement of certain natural objects with respect to one another, namely with parts of the earth, measuring lines, measuring rods, etc. It is clear that the system of concepts of axiomatic geometry alone cannot make any assertions as to the behavior of real objects of this kind, which we will call practically rigid bodies. To be able to make such assertions, geometry must be stripped of its merely logical-formal character by assigning to the empty conceptual schemata of axiomatic geometry objects of reality that are capable of being experienced.

To accomplish this, we need only add the proposition: Solid bodies[4] are related, with respect to their possible relative positions, as are bodies in Euclidean geometry of three dimensions. Then the propositions of Euclid contain assertions as to the behavior of practically rigid bodies.

Geometry thus amended is evidently a natural science; we may in fact regard it as the most ancient branch of physics. Its assertions rest essentially on induction from experience, but not on logical inferences only. We will call this amended geometry "practical geometry," and shall distinguish it in what follows from "purely axiomatic geometry." The question whether the practical geometry of the universe is Euclidean or not has a clear meaning, and its answer can only be furnished by experience. All linear measurement in physics is practical geometry in this sense, so too is geodetic and astronomical linear measurement, if we call to our help the law of experience that light is propagated in a straight line, and indeed in a straight line in the sense of practical geometry.

I attach special importance to the view of geometry I have just set forth, because without it I should have been unable to formulate the theory of relativity. Without it the following reflection would have been impossible: In a system of reference rotating relatively to an inertial system, the laws of relative position of rigid bodies do not correspond to the rules of Euclidean geometry on account of the Lorentz contraction; thus if we admit non-inertial systems as equally justified, we must abandon Euclidean geometry. The decisive step in the transition to general covariant equations would certainly not have been taken if the above interpretation had not served as a foundation. If one rejects the relation between the body of axiomatic Euclidean geometry and the practically rigid body of reality, one readily arrives at the following view, which was entertained in particular by that acute and profound thinker, H. Poincaré: Euclidean geometry is distinguished from all other imaginable axiomatic geometries by its simplicity. Now since axiomatic geometry *by itself* contains no assertions as to the reality that can be experienced, but can do so only in combination with physical laws, it should be possible and reasonable—whatever the nature of reality may be—to retain Euclidean geometry. For if contradictions between theory and experience manifest themselves, we should rather decide to change physical laws than to change axiomatic Euclidean geometry. If one rejects the relation between the practically rigid body and geometry, indeed one will not easily free oneself from the convention that Euclidean geometry is to be retained as the simplest. Why is the equivalence between the body that is practically rigid in experience and the body of geometry—which suggests itself so readily—denied by Poincaré and other researchers? Simply because under closer inspection the real solid bodies in nature are not rigid, because their geometrical behavior, that is, their potential relative positions (*Lagerungsmöglichkeiten*), depend upon temperature, external forces, etc. Thus the original, immediate relation between geometry and physical reality appears destroyed and one feels impelled toward the following more general view, which characterizes Poincaré's standpoint. Geometry (G) predicates nothing about the behavior of real things, but only geometry together with the totality (P) of physical laws can do so. Using symbols, we may say that only the sum of $(G) + (P)$ is subject to the test of experience. Thus (G) may be chosen arbitrarily, and also parts of (P); all these laws are conventions. All that is necessary to avoid contradictions is to choose the remainder of (P) so that (G) and the whole of (P) together are in accord with experience. According to this interpretation, axiomatic geometry and the part of natural law that has been given a conventional status appear as epistemologically equivalent.

Sub specie aeterni Poincaré is right in this interpretation, in my opinion.[5] The concept of the measuring-rod and the concept of the clock coordinated with it in the theory of relativity do not find their exact correspondence in the real world. It is also clear that the solid body and the clock do not play the part of irreducible elements in the conceptual edifice of physics, but that of composite constructs that must not be allowed to enter into the construction of theoretical physics as independent elements. But it is my conviction that in the present stage of development of theoretical physics these concepts must still be invoked as independent concepts, for we are still far from possessing such certain knowledge of theoretical principles as to be able to give exact theoretical constructions of such constructs [solid bodies and clocks].

Further, as to the objection that there are no really rigid bodies in nature, and that therefore the properties predicated of rigid bodies do not apply to physical reality, this objection is by no means so profound as might appear from a hasty examination. For it is not a difficult task to determine the physical state of a measuring body so accurately that its behavior in relation to the relative position of other measuring bodies will be sufficiently free from ambiguity to allow it to be substituted for the "rigid" body. It is to measuring bodies of this kind that statements as to rigid bodies must be referred.

All practical geometry is based upon a principle that is accessible to experience. To this principle we will now turn. We will call that which is enclosed between two boundaries, marked upon a practically rigid body, a line segment (*Strecke*). We imagine two practically rigid bodies, each with a line segment marked out on it. These two line segments are said to be "equal to one another" if the boundaries of one line segment can be brought to coincide permanently with the boundaries of the other. We now assume that:

If two line segments are found to be equal at one time and at some place, they are equal always and everywhere.

Not only the practical geometry of Euclid, but also its nearest generalization, the practical geometry of Riemann, and therewith the general theory of relativity, rest upon this assumption. Of the experimental reasons that warrant this assumption I will mention only one. The phenomenon of the propagation of light in empty space assigns a line segment, namely, the appropriate path of light, to each interval of local time, and vice versa. From this, it follows that the above assumption for line segments must also hold good for intervals of clock time in the theory of relativity. Consequently this assumption may then be formulated as follows: If two ideal clocks are going at the same rate at any time and at any place (being then in immediate proximity to each other), they will always go at the same rate, no matter where and when they are compared with each other at one place. If this law were not valid for natural clocks, the characteristic frequencies for the separate atoms of the same chemical element would not be in such exact agreement as experiment demonstrates.[6] The existence of sharp spectral lines is a convincing experimental proof of the above-mentioned basic principle of practical geometry. This is the ultimate foundation in fact which enables us to speak with meaning of the metric, in Riemann's sense of the word, of the four-dimensional space-time continuum.

The question whether the structure of this continuum is Euclidean, or in accordance with Riemann's general scheme, or otherwise, is, according to the view which is here being advocated, properly speaking a physical question that must be answered by experience, and not a question of a mere convention to be selected purely on grounds of expediency. Riemann's geometry will hold if the laws of relative position of practically rigid bodies approach those of the bodies of Euclid's geometry with an exactitude which increases in proportion as the dimensions of the part of space-time under consideration are diminished.

It is true that this proposed physical interpretation of geometry breaks down when applied immediately to spaces of sub-molecular order of magnitude. But nevertheless, even in questions as to the constitution of elementary particles, it retains part of its significance. For even when it is a question of describing the electrical elementary particles constituting matter, the attempt may still be made to ascribe physical importance to those concepts of fields that have been physically defined for the purpose of describing the geometrical behavior of bodies that are large as compared with the molecule. Only the outcome can decide the justification of such an attempt, which postulates physical reality for the fundamental principles of Riemann's geometry outside of the domain of their physical definitions. It might possibly turn out that this extrapolation has no better warrant than the extrapolation of the idea of temperature to the parts of a body of molecular order of magnitude.

It appears less problematic to extend the concepts of practical geometry to spaces of cosmic order of magnitude. It might, of course, be objected that a construction composed of solid rods departs more and more from ideal rigidity in proportion as its spatial extent becomes greater. But it will hardly be possible, I think, to assign fundamental significance to this objection. Therefore the question whether the universe is spatially finite or not seems to me decidedly an entirely meaningful question in the sense of practical geometry. I do not even consider it impossible that this question will be answered before long by astronomy. Let us call to mind what the general theory of relativity teaches in this respect. It offers two possibilities:

1. The universe is spatially infinite. This can be so only if the average spatial density of the matter in the universe, concentrated in the stars, vanishes, i.e., if the ratio of the total mass of the stars to the volume of the space through which they are scattered approaches zero as greater and greater volumes are considered.

2. The universe is spatially finite. This is only possible if there is a nonzero mean density of the ponderable matter in the universe. The smaller that mean density, the greater is the volume of the universe.

I must not fail to mention that a theoretical argument can be adduced in favor of the hypothesis of a finite universe. The general theory of relativity teaches that the inertia of a given body is greater the more ponderable masses are near it; thus it seems very natural to reduce the total effect of inertia of a body to interaction between it and the other bodies in the universe, as indeed, ever since Newton's time, gravity has been completely reduced to action and reaction between bodies. From the equations of the general theory of relativity it can be deduced that this total reduction of inertia to reciprocal action between masses—as required by E. Mach, for example—is possible only if the universe is spatially finite.[7]

This argument makes no impression on many physicists and astronomers. In the last analysis, experience alone can finally decide which of the two possibilities is realized in nature. How can experience furnish an answer? At first it might seem possible to determine the mean density of matter by observation of that part of the universe that is accessible to our perception. This hope is illusory. The distribution of the visible stars is extremely irregular, so that we in no way may venture to set down the mean density of star-matter in the universe as equal, let us say, to the mean density in the Milky Way. In any case, however great the space examined may be, one could not always surmise that there were no more stars beyond that space. So it seems impossible to estimate the mean density.

But there is another way, which seems to me more practicable, though it too presents great difficulties. For if we inquire into the deviations shown by the consequences of the general theory of relativity that are accessible to experience in the field of astronomy, when these are compared with the consequences of the Newtonian theory, we first of all find a deviation which shows itself in close proximity to gravitating mass and has been confirmed in the case of the planet Mercury. But if the universe is spatially finite there is a second deviation from the Newtonian theory, which, in the language of the Newtonian theory, may be expressed thus: The gravitational field is in its nature such as if it were produced not only by the ponderable masses, but also by a negative mass-density, distributed uniformly throughout space. Since this fictitious mass-density would have to be enormously small, it could make its presence felt only in very extended gravitating systems.[8]

Assuming that we know, let us say, the statistical distribution of the stars in the Milky Way, as well as their masses, then by Newton's law we can calculate the gravitational field and the mean velocities which the stars must have, so that the Milky Way should not collapse under the mutual attraction of its stars, but should maintain its actual extension. Now if the actual velocities of the stars, which can, of course, be measured, were smaller than the calculated velocities, we should have a proof that the actual attractions at great distances are smaller than by Newton's law. From such a deviation, one can prove indirectly that the universe is finite and even estimate its spatial magnitude.[9]

Can we imagine a picture of a three-dimensional universe that is finite, yet unbounded?

The usual answer to this question is "No," but that is the wrong answer. The purpose of the following remarks is to show why this is so. I want to show that without much difficulty we can illustrate the theory of a finite universe by means of a clear picture to which, with some practice, we shall soon grow accustomed.

First of all, an observation of epistemological nature. A geometrical-physical theory as such is at first necessarily incapable of being directly pictured, being purely a system of concepts. But these concepts serve the purpose of bringing a multiplicity of real or imaginary sensory experiences into connection in thought. To "visualize" a theory, therefore, means to give a representation to that abundance of experiences for which the theory supplies the schematic arrangement. In the present case, we have to ask: how could one can represent that behavior of solid bodies with respect to their reciprocal positions (contact) corresponding to the theory of a finite universe? There is really nothing new in what I have to say about this; but innumerable queries addressed to me prove that in this regard the needs of those eager for knowledge have not yet entirely been met. So will the initiated please pardon me if part of what I shall bring forward has long been known?

What do we wish to express when we say that our space is infinite? Nothing more than that we might lay any number whatever of bodies of equal sizes side by side without ever filling space. Suppose that we are provided with a great many cubical boxes all of the same size. In accordance with Euclidean geometry we can place them above, beside, and behind one another so as to fill a part of space as big as we like; but this construction would never be finished; we could go on adding more and more boxes without ever finding that there was no more room. That is what we wish to express when we say that space is infinite. It would be better to say that space is infinite in relation to practically rigid bodies, assuming that the laws of relative position for these bodies are given by Euclidean geometry.

Another example of an infinite continuum is the plane. On a plane surface we may adding squares of cardboard so that each side of any square has the side of another square adjacent to it. The construction is never finished; we can always go on laying squares if their laws of relative position correspond to those of plane figures of Euclidean geometry. The plane is therefore infinite in relation to the cardboard squares. Accordingly we say that the plane is an infinite continuum of two dimensions, and space an infinite continuum of three dimensions. What is here meant by the number of dimensions, I think I may assume to be known.

Now we take an example of a two-dimensional continuum that is finite, but unbounded. We imagine the surface of a large globe and a quantity of small paper discs, all of the same size. We place one of the discs anywhere on the surface of the globe. If we move the disc about with our fingers, anywhere we like on the surface of the globe, we do not come upon a boundary anywhere on the journey. Therefore we say that the spherical surface of the globe is an unbounded continuum. Moreover, the spherical surface is a finite continuum. For if we stick the paper discs on the globe, so that no disc overlaps another, the surface of the globe will finally become so full that there is no room for another disc. This simply means that the spherical surface of the globe is finite in relation to the paper discs. Further, the spherical surface is a non-Euclidean continuum of two dimensions, that is to say, the laws of relative position for the rigid figures lying in it do not agree with those of the Euclidean plane. This can be shown in the following way. Surround one paper disc by six discs placed in a circle around it, each of these surrounded again by six discs, and so on. If this construction is made on a plane surface, we have an uninterrupted arrangement in which there are six discs touching every disc except those which lie on the outside.

On the spherical surface the construction also seems to promise success at the outset, and the smaller the radius of the disc in proportion to that of the sphere, the more promising it seems. But as the construction progresses it becomes more

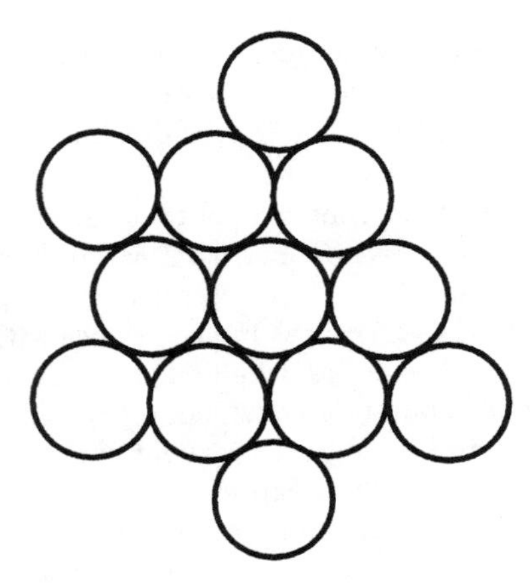

Figure 1

and more evident that the arrangement of the discs in the manner indicated is not possible without interruption, as it should be possible according to the Euclidean geometry of the plane surface. In this way, creatures who cannot leave the spherical surface and cannot even peep out from the spherical surface into three-dimensional space, might discover, merely by experimenting with discs, that their two-dimensional "space" is not Euclidean, but spherical.

From the latest results of the theory of relativity it is probable that our three-dimensional space is also approximately spherical, that is, that the laws of relative position of rigid bodies in it are not given by Euclidean geometry, but approximately by spherical geometry, if only we consider parts of space which are sufficiently large.[10] Now this is the place where the reader's imagination revolts. "Nobody can imagine this thing," he cries indignantly. "It can be said, but cannot be thought. I can think a spherical surface well enough, but nothing analogous to it in three dimensions."

We must try to surmount this obstacle in way of thought, and the patient reader will see that it is by no means a particularly difficult task. For this purpose, we will first give our attention once more to the geometry of two-dimensional spherical surfaces. In the adjoining figure let K be the spherical surface, touched at S by a plane, E, which, in order to help picture it, is shown in the drawing as a bounded surface. Further, let L be a disc on the spherical surface. Now let us imagine that at the point N of the spherical surface, diametrically opposite to S, there is attached a point source of light, throwing a shadow L' of the disc L upon the plane E. Every point on the sphere has its shadow on the plane. If the disc on the sphere K is moved, its shadow L' on the plane E also moves. When the disc L is at S, it almost exactly coincides with its shadow. If it moves on the spherical surface away from S upwards, the disc shadow L' on the plane also moves away from S on the plane outwards, growing ever larger. As the disc L approaches the point source of light N, the shadow moves off to infinity and becomes infinitely large.

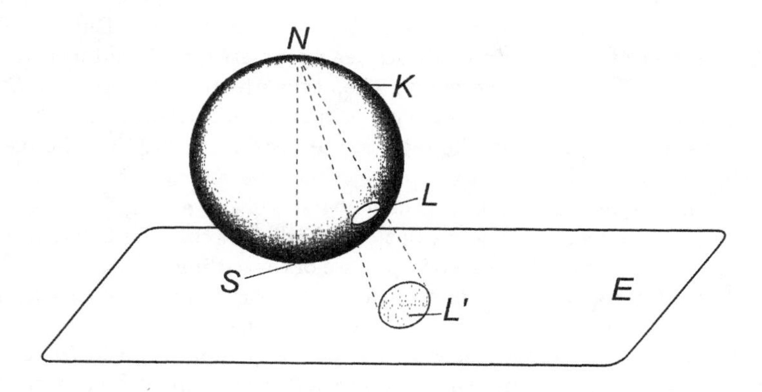

Figure 2

Now we ask: What are the laws of the relative positions of the disc-shadows L' on the plane E? Evidently they are exactly the same as the laws of the relative positions of the discs L on the spherical surface. For to each original figure on K there is a corresponding shadow figure on E. If two discs on K touch each other, their shadows on E also touch. The shadow-geometry on the plane agrees with the disc-geometry on the sphere. If we call the disc-shadows rigid figures, then spherical geometry holds good on the plane E with respect to these rigid figures. In particular, the plane is finite with respect to the disc-shadows, since only a finite number of the shadows can find room on the plane.

At this point somebody will say, "That is nonsense. The disc-shadows are precisely *not* rigid figures. We have only to move a measuring rod about on the plane E to convince ourselves that the shadows constantly increase in size as they move away from S on the plane towards infinity." But what if the measuring rods were to behave on the plane E in the same way as the disc-shadows L'? Then it would be impossible to show that the shadows increase in size as they move away from S; such an assertion would then no longer have any meaning whatever. In fact, the only objective assertion that can be made about the disc-shadows is just this, that, geoemtrically speaking, they behave just like rigid discs on the spherical surface in the sense of Euclidean geometry.

We must carefully bear in mind that our statement as to the growth of the disc-shadows as they move away from S towards infinity has no objective meaning in itself, as long as we are unable to employ Euclidean rigid bodies which can be moved about on the plane E for the purpose of comparing the size of the disc-shadows. In respect of the laws of relative position of the shadows L', the point S has no special privileges on the plane any more than on the spherical surface.

The illustration given above of spherical geometry on the plane is important for us because it readily allows itself to be transferred to the three-dimensional case.

Let us imagine a point S of our space, and a great number of small spheres, L', which can all be brought to coincide with one another. But these spheres are not to be rigid in the sense of Euclidean geometry; their radius is to increase (in the sense of Euclidean geometry) when they are moved away from S towards infinity, and this increase is to take place in exact accordance with the same law as applies to the increase of the radii of the disc-shadows L' on the plane.

After having gained a vivid image of the geometrical behavior of our L' spheres, let us assume that in our space there are no rigid bodies at all in the sense of Euclidean geometry, but only bodies having the behavior of our L' spheres. Then we shall have a vivid picture of three-dimensional spherical space, or, rather of three-dimensional spherical geometry. Here our spheres must be called "rigid" spheres. Their increase in size as they depart from S is not to be detected by measuring with measuring rods, any more than in the case of the disc-shadows on E, because the standards of measurement behave in the same way as the spheres. Space is homogeneous, that is to say, the same spherical configurations are possible in the environment of all points.* Our space is finite, because, in consequence of the "growth" of the spheres, only a finite number of them can find room in space.

In this way, by using as a crutch the practice in thinking and visualization that Euclidean geometry gives us, we have acquired a clear picture of spherical geometry. We may without difficulty impart more depth and vigor to these ideas by carrying out special imaginary constructions. Nor would it be difficult to illustrate the case of what is called elliptical geometry in an analogous manner.[11] My only aim today has been to show that the human faculty of visualization is by no means bound to capitulate to non-Euclidean geometry.

Notes

[Originally delivered as a lecture on January 27, 1921; expanded (as noted below) when published separately (Einstein 1921, *ECP* 7:382–405); this is a revision and correction of the translation by Sonja Bargmann in Einstein 1982, 231–246.]

1. [See, for instance, Hilbert 1959.]
2. [See Schlick 1963 and Friedman 2002, 203–206.]
3. [See Friedman 2001 and his excellent commentary in Friedman 2002.]
4. [In this and his other essays in this anthology, Einstein seems to make a clear distinction between *feste Körper* and *starrer Körper*, which I have accordingly consistently rendered as "solid bodies" and "rigid bodies," respectively.

*This is intelligible without calculation—albeit only for the two-dimensional case—thus going back once more to the case of the disc on the surface of the sphere.

This is significantly different from Helmholtz's usage, as noted above, 51, note 2. In my opinion, Einstein's usage reflects his awareness of the problem of the failure of rigidity already in special relativity.]

5. [Einstein's turn of phrase here is noticeably reminiscent of Spinoza's phrase *sub specie aeternitatis*, "from the point of view of eternity," as in his *Ethics*, Part II, Proposition 44, Corollary 2: "It is of the nature of reason to perceive things under a certain species of eternity." For Einstein's relation to Poincaré in the present essay, see Paty 1992 and Friedman 2002, 198; for Einstein's relation to Spinoza, see Paty 1986 and Pesic 1996.]

6. [By "characteristic frequencies" (*Eigenfrequenzen*) Einstein is referring to the discrete and unique pattern of sharp spectral lines each element always shows when heated to incandescence.]

7. [Einstein long believed that general relativity was consistent with (or even required) Mach's principle (as enunciated here). But later studies showed, on the contrary, that Mach's principle has no necessary connection with general relativity, as Einstein himself came to realize in his later years. In 1954, he wrote a friend that "one should not longer speak of Mach's principle." See Pais 1982, 284–288. For Mach's own sympathetic response to Riemann, see Mach 1906.]

8. [This is the famous cosmological constant, which Einstein introduced into the field equations of general relativity first of all in order to give a static solution (as he thought necessary at the time) but secondly because such a term (expressed by a scalar constant Λ multiplied by the metric tensor, $\Lambda g_{\mu\nu}$) must in general be included beside the contracted Riemann tensor ($R_{\mu\nu}$). The observational discovery around the beginning of the twenty-first century that this constant was nonzero but small ($\Lambda \approx 0.7$) represented a great theoretical challenge, the problem of "dark energy." For Einstein's 1917 introduction of the cosmological term, see *ECP* 6:540–552, also available in Lorentz et al. 1923, 177–188.]

9. [Here the text of Einstein's original 1921 address ended; the remainder of the essay as given here follows his published text in Einstein 1921.]

10. [Einstein argued for such a spherical model in his 1917 paper cited in note 7.]

11. [In 1917, Einstein had corresponded with Erwin Freundlich and Felix Klein about the possibility of elliptical cosmological solutions of the equations of general relativity; see *ECP* 8:393–394, 425–426 [287, 311].]

Non-Euclidean Geometry and Physics (1925)

Albert Einstein

Consideration of the relations of non-Euclidean geometry to physics leads necessarily to the problem of the relation between geometry and physics in general. I will address this problem first and will try to keep clear of controversial questions of philosophy.

In the most ancient times, geometry was doubtless a half-empirical science, a kind of primitive physics. A point was a body whose extension was ignored; a straight line was defined, for instance, through points that could be rendered optically coincident along the line of sight, or through a string pulled tight.[1] It is thus a matter of concepts that indeed—as is always the case with concepts—are not derived simply from experience, that is, not logically inferred from experience but nevertheless are directly put in relation to things of experience. At this stage of knowledge, propositions about points, straight lines, and the equality of distances and angles were at the same time the knowledge of propositions about certain experiences with natural objects.

Understood in this way, geometry became a mathematical science in that one realized that most of its propositions could be derived from a few of them, the so-called axioms, in a purely logical way. For every science that concerns itself solely with *logical* relations between given objects according to given rules is mathematics. All attention is now focused on inferring relations, for the independent construction of a logical system—uninfluenced by uncertain external experiences subject to chance—has always had an irresistible fascination for the human mind.

In the system of geometry, only the fundamental ideas of point, straight line, distance, etc., and also the so-called axioms respectively remain logically irreducible as testimony to its empirical provenance. One sought to limit the number of these logically irreducible fundamental ideas and axioms to a minimum. The endeavor to raise geometry out of the murky sphere of the empirical now led unnoticed to a intellectual transposition, somewhat analogous to the promotion of the revered heroes of antiquity to gods. For one gradually became accustomed to regard the basic ideas and axioms of geometry as "self-evident," that is, as objects and qualities of representation present as such in the human spirit, in such a way that the fundamental concepts of geometry conform to objects of inner intuition and that a negation of an axiom of geometry cannot at all be carried out meaningfully. Given this attitude, the applicability of these fundamental tenets to the objects of reality then becomes a problem, and we may very well add that it is the same problem from which Kant's conception of space arose.

Physics provided a second motive for the separation of geometry from its empirical basis. According to its refined conception of the nature of solid bodies and of light, there are no natural objects whose characteristics *exactly* accord with the basic concepts of Euclidean geometry.[2] A solid body is not inflexible and a light ray does not exactly embody a straight line, nor indeed any one-dimensional structure. According to modern science, geometry by itself does not agree with any experiences, strictly speaking, but only geometry taken together with mechanics, optics, etc. Moreover, because geometry must precede physics in that the laws of the latter cannot be expressed without those of the former, geometry appears to be a science that logically precedes every experience and every experimental science. Thus it came to be that at the beginning of the nineteenth century, the foundation of Euclidean geometry appeared to be something absolutely unshakeable not only to mathematicians and philosophers but also to physicists.

In addition, one can say that throughout the nineteenth century, the situation presented itself still more crudely, schematically, and rigidly to the physicist, provided that he did not directly pay attention to the theory of knowledge,. His unconsciously held viewpoint corresponded to both propositions: the concepts and fundamental propositions of Euclidean geometry are self-evident. If certain conditions are met, designated rigid bodies will realize the geometrical concept of distance and light rays will realize the concept of straight lines.

The overcoming of this position was a difficult piece of work that took about a century. Strangely enough, it had its origin in purely mathematical investigations, long before the garment of Euclidean geometry became too tight for physics. It belongs to the tasks of mathematicians to ground geometry on the minimum of axioms. Among Euclid's axioms there was one that appeared to mathematicians less immediately intelligible than the others. They therefore strove for a long time to reduce it to the others, that is, to prove it from them. This was the so-called parallel postulate. Because all the pains taken to achieve such a proof came to nothing, the conjecture had gradually to emerge that such a proof is impossible, that is, that this postulate is independent from the others. This could be proved, in that one could erect an edifice free from logical contradiction, which only differs from Euclidean geometry by replacing the parallel postulate with another. It remains the imperishable merit of Lobachevsky and of Bolyai (father and son) that they independently grasped and convincingly realized these thoughts.[3]

Thus, the conviction had to become entrenched among mathematicians that, beside Euclidean geometry exist yet other geometries, logically equal in rights. One could not fail to ask whether physics necessarily had to be based on Euclidean geometry and no other. One might put the question in the most decisive form: is Euclidean or some other geometry valid in the physical world?

Whether this latter question is meaningful has been much disputed. In order to see the matter clearly, one must consistently adopt one of two points of view. In the first, one holds that the "body" of geometry is realized in principle by rigid bodies in nature, provided that certain conditions are met regarding temperature, mechanical strain, etc.; this is the point of view of the practical physicist.[4] In this case, the "distance" of geometry agrees with a natural object and thereby all propositions of geometry gain the character of assertions about real bodies. This point of view was especially clearly advocated by Helmholtz, and we can add that without him the formulation of relativity theory would have been practically impossible.

In the other point of view, one denies in principle the existence of objects that agree with the fundamental concepts of geometry. Then geometry by itself would include no assertions about objects of reality, only geometry taken together with physics. This point of view, which may more complete for the systematic representation of a finished physics, was expounded particularly clearly by Poincaré. From this standpoint, the entire content of geometry is conventional; which geometry is preferable depends on how "simple" physics can be made by using geometry to agree with experience.

Here, we wish to choose the first point of view as better suited to the present situation of our knowledge. Seen from this standpoint, our question about the validity or invalidity of Euclidean geometry has a clear meaning. Euclidean geometry or geometry in general retains now, as before, the character of a mathematical science, in that the derivation of its propositions from its axioms remains purely logical, but at the same time it becomes a physical science in that its axioms include assertions about objects in nature, assertions whose correctness only experiment can decide. But we must always be cognizant of the fact that the idealization that lies in the fiction of a rigid (measuring) body as a natural object might someday be shown to be unjustified or else justified only in relation to certain natural phenomena. General relativity has already shown the illegitimacy of this concept for spaces that are not small in the astronomical sense. The theory of electrical elementary quanta could show the illegitimacy of this concept for distances of the atomic order of magnitude. Riemann already knew both to be possible.

Riemann's merit for the development of our ideas about the connection between geometry and physics is twofold. First, he discovered the spherical-elliptical geometry, the counterpart to Lobachevsky's hyperbolic geometry. Thus for the first time he showed the possibility that geometrical space might have a finite extension in a metrical sense. This idea was immediately understood and has led to the often-considered question whether physical space is not finite.

Second, Riemann had the bold idea to create a geometry that is incomparably more general than Euclid's or than the non-Euclidean geometries in the narrower sense. He thus created "Riemannian geometry," which (like non-Euclidean geometries in the narrower sense) is Euclidean only in the infinitely small; it transfers the Gaussian theory of surfaces to a continuum of arbitrarily many dimensions. According to this more generalized geometry, the metrical properties of space or the possible disposition of infinitely many, infinitely small rigid bodies within finite regions is not fixed by the axioms of geometry alone. Instead of being discouraged by this insight or inferring from it the physical meaninglessness of his system, Riemann had the bold idea that the geometrical behavior of bodies might depend on physical realities or forces. Thus, through purely mathematical speculations he came to the idea of the inseparability of geometry from physics, an idea that, seventy years later, gained acceptance indeed in general relativity theory, where geometry and gravitational theory merged into a unity.

After Riemannian geometry was brought into its simplest form by Levi-Civita through the introduction of the concept of infinitesimal parallel displacement, Riemannian geometry was even further generalized by Weyl and Eddington in the hope that in such a broadened conceptual system the electromagnetic law might also find a place.[5] Whatever the outcome of these efforts might be, in any case one is right to say: the ideas that have developed from non-Euclidean geometry have proved to be eminently fruitful for modern theoretical physics.

Notes

[Originally appeared in Einstein 1925.]

1. [Note the resemblance between this image and that of Poincaré 1898, 140 above.]

2. [Throughout I have translated Einstein's term *fester Körper* as "solid body," in contrast to his distinct uses of another term, *starrer Körper*, to designate a "rigid body." In contrast, Einstein's usage seems to differ from that of Helmholtz (for instance), who only uses the one adjective *fester* to mean "rigid," which I kept to in the translation of the Helmholtz essays above. Einstein doubtless became aware of the complexities of defining "rigidity" in the context of his work on special relativity, in which absolute rigidity proved impossible because any solid body cannot respond simultaneously to an impact (say) at some point on the body simply because the impulse requires time to traverse the body.]

3. [Bolyai's father Wolfgang had been a close friend of Gauss and had also studied closely the problem of the parallel postulate; indeed, he advised his son János against his own obsession with the problem: "You must not attempt this approach to parallels: I know this way to its very end. I have traversed this bottomless night, which extinguished all light and joy of my life. For God's sake! I entreat you to leave parallels alone, abhor them like indecent talk, they may deprive you from your time, health, tranquility, and the happiness of your life." See Prékopa 2006, 16, Gray 2004 and 2006.]

4. [For instance, the temperature must be held constant lest the body expand or contract thermally.]

5. [For Levi-Civita, see Bottazzini 1999; for Weyl, see Scholz 1999.]

Space-Time (1926)

Albert Einstein

All our thoughts and concepts are called up by sense-experiences and have a meaning only in reference to these sense-experiences. On the other hand, however, they are products of the spontaneous activity of our minds; they are thus in no wise logical consequences of the contents of these sense-experiences. If, therefore, we wish to grasp the essence of a complex of abstract notions we must for the one part investigate the mutual relationships between the concepts and the assertions made about them; for the other, we must investigate how they are related to the experiences.

So far as the way is concerned in which concepts are connected with one another and with the experiences there is no difference of principle between the concept-systems of science and those of daily life. The concept-systems of science have grown out of those of daily life and have been modified and completed according to the objects and purposes of the science in question.

The more universal a concept is the more frequently it enters into our thinking; and the more indirect its relation to sense-experience, the more difficult it is for us to comprehend its meaning; this is particularly the case with pre-scientific concepts that we have been accustomed to use since childhood. Consider the concepts referred to in the words "where," "when," "why," "being," to the elucidation of which innumerable volumes of philosophy have been devoted. We fare no better in our speculations than a fish which should strive to become clear as to what is water.

Space

In the present article we are concerned with the meaning of "where," that is, of space. It appears that there is no quality contained in our individual primitive sense-experiences that may be designated as spatial. Rather, what is spatial appears to be a sort of order of the material objects of experience.[1] The concept "material object" must therefore be available if concepts concerning space are to be possible. It is the logically primary concept. This is easily seen if we analyze the spatial concepts for example, "next to," "touch," and so forth, that is, if we strive to become aware of their equivalents in experience. The concept "object" is a means of taking into account the persistence in time or the continuity, respectively, of certain groups of experience-complexes. The existence of objects is thus of a conceptual nature, and the meaning of the concepts of objects depends wholly on their being connected (intuitively) with groups of elementary sense-experiences. This connection is the basis of the illusion which makes primitive experience appear to inform us directly about the relation of material bodies (which exist, after all, only in so far as they are thought).

In the sense thus indicated we have (the indirect) experience of the contact of two bodies. We need do no more than call attention to this, as we gain nothing for our present purpose by singling out the individual experiences to which this assertion alludes. Many bodies can be brought into permanent contact with one another in manifold ways. We speak in this sense of the position-relationships of bodies (*Lagenbeziehungen*). The general laws of such position-relationships are essentially the concern of geometry. This holds, at least, if we do not wish to restrict ourselves to regarding the propositions that occur in this branch of knowledge merely as relationships between empty words that have been set up according to certain principles.

Pre-scientific Thought.—Now, what is the meaning of the concept "space" which we also encounter in pre-scientific thought? The concept of space in pre-scientific thought is characterised by the sentence: "we can think away things but not the space which they occupy." It is as if, without having had experience of any sort, we had a concept, nay even a presentation, of space and as if we ordered our sense-experiences with the help of this concept, present *a priori.* On the other hand, space appears as a physical reality, as a thing which exists independently of our thought, like material objects. Under the influence of this view of space the fundamental concepts of geometry: the point, the straight line, the plane, were even regarded as having a self-evident character. The fundamental principles that deal with these configurations were regarded as being necessarily valid and as having at the same time an objective content. No scruples were felt about ascribing an objective meaning to such statements as "three empirically given bodies (practically infinitely small) lie on one straight line," without demanding a physical definition for such an assertion. This blind faith in evidence and in the immediately real meaning of the concepts and propositions of geometry became uncertain only after non-Euclidean geometry had been introduced.

Reference to the Earth.—If we start from the view that all spatial concepts are related to contact-experiences of solid bodies, it is easy to understand how the concept "space" originated, namely, how a thing independent of bodies and yet embodying their position-possibilities (*Lagerungsmöglichkeiten*) was posited. If we have a system of bodies in contact and at rest relatively to one another, some can be replaced by others. This property of allowing substitution is interpreted as "available space." Space denotes the property in virtue of which rigid bodies can occupy different positions. The view that space is something with a unity of its own is perhaps due to the circumstance that in pre-scientific thought all positions of bodies were referred to one body (reference body), namely the earth. In scientific thought the earth is represented by the co-ordinate system. The assertion that it would be possible to place an unlimited number of bodies next to one another denotes that space is infinite. In pre-scientific thought the concepts "space" and "time" and "body of reference" are scarcely differentiated at all. A place or point in space is always taken to mean a material point on a body of reference.

Euclidean Geometry.—If we consider Euclidean geometry we clearly discern that it refers to the laws regulating the positions of rigid bodies. It turns to account the ingenious thought of tracing back all relations concerning bodies and their relative positions to the very simple concept "distance" (*Strecke*). Distance denotes a rigid body on which two material points (marks) have been specified. The concept of the equality of distances (and angles) refers to experiments involving coincidences; the same remarks apply to the theorems on congruence. Now, Euclidean geometry, in the form in which it has been handed down to us from Euclid, uses the fundamental concepts "straight line" and "plane" which do not appear to correspond, or at any rate, not so directly, with experiences concerning the position of rigid bodies. On this it must be remarked that the concept of the straight line may be reduced to that of the distance.* Moreover, geometricians were less concerned with bringing out the relation of their fundamental concepts to experience than with deducing logically the geometrical propositions from a few axioms enunciated at the outset.

Let us outline briefly how perhaps the basis of Euclidean geometry may be gained from the concept of distance.

We start from the equality of distances (axiom of the equality of distances). Suppose that of two unequal distances one is always greater than the other. The same axioms are to hold for the inequality of distances as hold for the inequality of numbers.

Three distances $\overline{AB'}$, $\overline{BC'}$, $\overline{CA'}$ may, if CA' be suitably chosen, have their marks BB', CC', AA' superposed on one another in such a way that a triangle ABC results. The distance CA' has an upper limit for which this construction is still just possible. The points $A, (BB')$ and C then lie in a "straight line" (definition). This leads to the concepts: producing a distance by an amount equal to itself; dividing a distance into equal parts; expressing a distance in terms of a number by means of a measuring-rod (definition of the space-interval between two points).

When the concept of the interval between two points or the length of a distance has been gained in this way we require only the following axiom (Pythagoras' theorem) in order to arrive at Euclidean geometry analytically.

To every point of space (body of reference) three numbers (co-ordinates) x, y, z may be assigned—and conversely—in such a way that for each pair of points $A(x_1, y_1, z_1)$ and $B(x_2, y_2, z_2)$ the theorem holds:

$$\textit{measure-number } AB = \sqrt{(x_2 - x_1)^2 + (y_2 - y_1)^2 + (z_2 - z_1)^2}.$$

*A hint of this is contained in the theorem: "the straight line is the shortest connection between two points." This theorem served well as a definition of the straight line, although the definition played no part in the logical texture of the deductions.

All further concepts and propositions of Euclidean geometry can then be built up purely logically on this basis, in particular also the propositions about the straight line and the plane.

These remarks are not, of course, intended to replace the strictly axiomatic construction of Euclidean geometry. We merely wish to indicate plausibly how all conceptions of geometry may be traced back to that of distance. We might equally well have epitomized the whole basis of Euclidean geometry in the last theorem above. The relation to the foundations of experience would then be furnished by means of a supplementary theorem.

The co-ordinate may and *must* be chosen so that two pairs of points separated by equal intervals, as calculated by the help of Pythagoras' theorem, may be made to coincide with one and the same suitably chosen distance (on a solid).

The concepts and propositions of Euclidean geometry may be derived from Pythagoras' proposition without the introduction of rigid bodies; but these concepts and propositions would not then have contents that could be tested. They are not "true" propositions but only logically correct propositions of purely formal content.

Difficulties.—A serious difficulty is encountered in the above represented interpretation of geometry in that the rigid body of experience does not correspond *exactly* with the geometrical body. In stating this I am thinking less of the fact that there are no absolutely definite marks than that temperature, pressure and other circumstances modify the laws relating to position. It is also to be recollected that the structural constituents of matter (such as atom and electron, *q.v.*[2]) assumed by physics are not in principle commensurate with rigid bodies, but that nevertheless the concepts of geometry are applied to them and to their parts. For this reason consistent thinkers have been disinclined to allow real contents of facts (*reale Tatsachenbestände*) to correspond to geometry alone.[3] They considered it preferable to allow the content of experience (*Erfahrungsbestände*) to correspond to geometry and physics conjointly.

This view is certainly less open to attack than the one represented above; as opposed to the atomic theory it is the only one that can be consistently carried through. Nevertheless, in the opinion of the author it would not be advisable to give up the first view, from which geometry derives its origin. This connection is essentially founded on the belief that the ideal rigid body is an abstraction that is well rooted in the laws of nature.[4]

Foundations of Geometry.—We come now to the question: what is *a priori* certain or necessary, respectively in geometry (doctrine of space) or its foundations? Formerly we thought everything—yes, everything; nowadays we think—nothing. Already the distance-concept is logically arbitrary; there need be no things that correspond to it, even approximately. Something similar may be said of the concepts straight line, plane, of three-dimensionality and of the validity of Pythagoras' theorem. Nay, even the continuum-doctrine is in no wise given with the nature of human thought, so that from the epistemological point of view no greater authority attaches to the purely topological relations than to the others.

Earlier Physical Concepts.—We have yet to deal with those modifications in the space-concept, which have accompanied the advent of the theory of relativity. For this purpose we must consider the space-concept of the earlier physics from a point of view different from that above. If we apply the theorem of Pythagoras to infinitely near points, it reads

$$ds^2 = dx^2 + dy^2 + dz^2,$$

where ds denotes the measurable interval between them. For an empirically-given ds the co-ordinate system is not yet fully determined for every combination of points by this equation. Besides being translated, a co-ordinate system may also be rotated.† This signifies analytically: the relations of Euclidean geometry are covariant with respect to linear orthogonal transformations of the co-ordinates.[5]

In applying Euclidean geometry to pre-relativistic mechanics a further indeterminateness enters through the choice of the co-ordinate system: the state of motion of the co-ordinate system is arbitrary to a certain degree, namely, in that substitutions of the co-ordinates of the form

$$\begin{aligned} x' &= x - vt \\ y' &= y \\ z' &= z \end{aligned}$$

also appear possible.[6] On the other hand, earlier mechanics did not allow co-ordinate systems to be applied of which the states of motion were different from those expressed in these equations. In this sense we speak of "inertial systems." In these favored-inertial systems we are confronted with a new property of space so far as geometrical relations are concerned.[7] Regarded more accurately, this is not a property of space alone but of the four-dimensional continuum consisting of time and space conjointly.

Appearance of Time.—At this point time enters explicitly into our discussion for the first time. In their applications space (place) and time always occur together. Every event that happens in the world is determined by the space-co-ordinates x, y, z, and the time-co-ordinate t. Thus the physical description was four-dimensional right from the beginning. But this four-dimensional continuum seemed to resolve itself into the three-dimensional continuum of space and the one-dimensional continuum of time. This apparent resolution owed its origin to the illusion that the meaning of the concept "simultaneity" is self-evident, and this illusion arises from the fact that we receive news of near events almost instantaneously owing to the agency of light.

This faith in the absolute significance of simultaneity was destroyed by the law regulating the propagation of light in empty space or, respectively, by the

†Change of direction of the co-ordinate axes while their orthogonality is preserved.

Maxwell-Lorentz electrodynamics. Two infinitely near points can be connected by means of a light-signal if the relation

$$ds^2 = c^2dt^2 - dx^2 - dy^2 - dz^2 = 0$$

holds for them.[8] It further follows that ds has a value which, for arbitrarily chosen infinitely near space-time points, is independent of the particular inertial system selected. In agreement with this we find that for passing from one inertial system to another, linear equations of transformation hold which do not in general leave the time-values of the events unchanged. It thus became manifest that the four-dimensional continuum of space cannot be split up into a time-continuum and a space-continuum except in an arbitrary way. This invariant quantity ds may be measured by means of measuring-rods and clocks.

Four-Dimensional Geometry.—On the invariant ds a four-dimensional geometry may be built up which is in a large measure analogous to Euclidean geometry in three dimensions. In this way physics becomes a sort of statics in a four-dimensional continuum. Apart from the difference in the number of dimensions the latter continuum is distinguished from that of Euclidean geometry in that ds^2 may be greater or less than zero. Corresponding to this we differentiate between time-like and space-like line-elements. The boundary between them is marked out by the element of the "light-cone" $ds^2 = 0$ which starts out from every point. If we consider only elements which belong to the same time-value, we have

$$-ds^2 = dx^2 + dy^2 + dz^2.$$

These elements ds may have real counterparts in distances at rest and, as before, Euclidean geometry holds for these elements.

Effect of Relativity, Special and General.—This is the modification which the doctrine of space and time has undergone through the restricted theory of relativity. The doctrine of space has been still further modified by the general theory of relativity, because this theory denies that the three-dimensional spatial section of the space-time continuum is Euclidean in character. Therefore it asserts that Euclidean geometry does not hold for the relative positions of bodies that are continuously in contact.

For the empirical law of the equality of inertial and gravitational mass led us to interpret the state of the continuum, in so far as it manifests itself with reference to a non-inertial system, as a gravitational field and to treat non-inertial systems as equivalent to inertial systems. Referred to such a system, which is connected with the inertial system by a non-linear transformation of the co-ordinates, the metrical invariant ds^2 assumes the general form:

$$ds^2 = \sum_{\mu\nu} g_{\mu\nu} dx^\mu dx^\nu,$$

where the $g_{\mu\nu}$'s are functions of the co-ordinates and where the sum is to be taken over the indices for all combinations 11, 12, ..., 44. The variability of the

$g_{\mu\nu}$'s is equivalent to the existence of a gravitational field. If the gravitational field is sufficiently general it is not possible at all to find an inertial system, that is, a co-ordinate system with reference to which ds^2 may be expressed in the simple form given above:

$$ds^2 = c^2dt^2 - dx^2 - dy^2 - dz^2;$$

But in this case, too, there is in the infinitesimal neighborhood of a space-time point a local system of reference for which the last-mentioned simple form for ds holds.

This state of the facts leads to a type of geometry which Riemann's genius created more than half a century before the advent of the general theory of relativity of which Riemann divined the high importance for physics.

Riemann's Geometry.—Riemann's geometry of an n-dimensional space bears the same relation to Euclidean geometry of an n-dimensional space as the general geometry of curved surfaces bears to the geometry of the plane. For the infinitesimal neighbourhood of a point on a curved surface there is a local co-ordinate system in which the distance ds between two infinitely near points is given by the equation

$$ds^2 = dx^2 + dy^2.$$

For any arbitrary (Gaussian) co-ordinate-system, however, an expression of the form

$$ds^2 = g_{11}dx_1^2 + 2g_{12}dx_1dx_2 + g_{22}dx_2^2$$

holds in a finite region of the curved surface. If the $g_{\mu\nu}$'s are given as functions of x_1 and x_2 the surface is then fully determined geometrically. For from this formula we can calculate for every combination of two infinitely near points on the surface the length ds of the minute rod connecting them; and with the help of this formula all networks that can be constructed on the surface with these little rods can be calculated. In particular, the "curvature" at every point of the surface can be calculated; this is the quantity that expresses to what extent and in what way the laws regulating the positions of the minute rods in the immediate vicinity of the point under consideration deviate from those of the geometry of the plane.

This theory of surfaces by Gauss has been extended by Riemann to continua of any arbitrary number of dimensions and has thus paved the way for the general theory of relativity. For it was shown above that corresponding to two infinitely near space-time points there is a number ds which can be obtained by measurement with rigid measuring-rods and clocks (in the case of time-like elements, indeed, with a clock alone). This quantity occurs in the mathematical theory in place of the length of the minute rods in three-dimensional geometry. The curves for which $\int ds$ has stationary values determine the paths of material points and rays of light in the gravitational field, and the "curvature" of space is dependent on the matter distributed over space.

Just as in Euclidean geometry the space-concept refers to the position-possibilities of rigid bodies, so in the general theory of relativity the space-time-concept refers to the behavior of rigid bodies and clocks. But the space-time-continuum differs from the space-continuum in that the laws regulating the behavior of these objects (clocks and measuring-rods) depend on where they happen to be. The continuum (or the quantities that describe it) enters explicitly into the laws of nature, and conversely these properties of the continuum are determined by physical factors. The relations that connect space and time can no longer be kept distinct from physics proper. Nothing certain is known of what the properties of the space-time-continuum may be as a whole. Through the general theory of relativity, however, the view that the continuum is infinite in its time-like extent but finite in its space-like extent has gained in probability.

Time

The physical time-concept answers to the time-concept of the extra-scientific mind. Now, the latter has its root in the time-order of the experiences of the individual, and this order we must accept as something primarily given.

I experience the moment "now," or, expressed more accurately, the present sense-experience (*Sinnen-Erlebnis*) combined with the recollection of (earlier) sense-experiences. That is why the sense-experiences seem to form a series, namely the time-series indicated by "earlier" and "later." The experience-series is thought of as a one-dimensional continuum. Experience-series can repeat themselves and can then be recognized. They can also be repeated inexactly, wherein some events are replaced by others without the character of the repetition becoming lost for us. In this way we form the time-concept as a one-dimensional frame which can be filled in by experiences in various ways. The same series of experiences answer to the same subjective time-intervals.

The transition from this "subjective" time (*Ich-Zeit*) to the time-concept of pre-scientific thought is connected with the formation of the idea that there is a real external world independent of the subject. In this sense the (objective) event is made to correspond with the subjective experience. In the same sense there is attributed to the "subjective" time of the experience a "time" of the corresponding "objective" event. In contrast with experiences external events and their order in time claim validity for all subjects.

This process of objectification would encounter no difficulties were the time-order of the experiences corresponding to a series of external events the same for all individuals. In the case of the immediate visual perceptions of our daily lives, this correspondence is exact. That is why the idea that there is an objective time-order became established to an extraordinary extent. In working out the idea of an objective world of external events in greater detail, it was found necessary to make events and experiences depend on each other in a more complicated way. This was at first done by means of rules and modes of thought instinctively gained, in which the conception of space plays a particularly prominent part. This process of refinement leads ultimately to natural science.

The measurement of time is effected by means of clocks. A clock is a thing which automatically passes in succession through a (practically) equal series of events (period). The number of periods (clock-time) elapsed serves as a measure of time. The meaning of this definition is at once clear if the event occurs in the immediate vicinity of the clock in space; for all observers then observe the same clock-time simultaneously with the event (by means of the eye) independently of their position. Until the theory of relativity was propounded it was assumed that the conception of simultaneity had an absolute objective meaning also for events separated in space.

This assumption was demolished by the discovery of the law of propagation of light. For if the velocity of light in empty space is to be a quantity that is independent of the choice (or, respectively, of the state of motion) of the inertial system to which it is referred, no absolute meaning can be assigned to the conception of the simultaneity of events that occur at points separated by a distance in space. Rather, a special time must be allocated to every inertial system. If no co-ordinate system (inertial system) is used as a basis of reference there is no sense in asserting that events at different points in space occur simultaneously. It is in consequence of this that space and time are welded together into a uniform four-dimensional continuum.

Notes

[First appeared in the *Encyclopædia Britannica* Thirteenth Edition (1926), here reproduced from the Fourteenth Edition (1929), vol. 21, 105–108, without change (except for regularizing the spelling), there being no German original available (except for partial manuscript drafts in the Einstein Archive). In the Fourteenth (1929) and subsequent editions (until 1939), there was appended to this article by Einstein a further section entitled "Electro-magnetic Field" by Arthur Stanley Eddington, describing the attempts (as of 1929) to make a unified field theory incorporating electromagnetism into the geometric structure of the theory, not merely as part of the stress-energy tensor. Eddington includes Weyl's 1918 theory, his own 1921 attempt (using affine geometry), and Einstein's 1929 theory using "distant parallelism."]

1. [Here Einstein may be thinking about the work of Jean Piaget, who had followed Einstein's suggestion to explore the development of spatial thinking in children and had found that children are not born with "ordinary" spatial perceptions but have to develop them over a long time. See Piaget 1971, 10, 82, 100, which discusses Einstein's question to Piaget whether the development of time-perception precedes or follows speed-perception in children.]

2. [This refers to articles on atom and electron (not by Einstein) in the *Encyclopædia Britannica.*]

3. [Without naming him, Einstein may well be alluding here to Helmholtz, especially in view of the reference to Helmholtz's preferred term "facts" (*Tatsachen*) as a description of the foundations of geometry.]

4. [Note here Einstein's hesitation about the validity of rigid bodies; curiously, he does not mention the most glaring difficulty raised by relativity theory itself, as discussed in the Introduction, that the finite time taken by all influences means that completely rigid bodies are hence impossible.]

5. [By "covariant," Einstein means that, however one rotates or translates Euclidean coordinates, the distance between two points remains invariant; thus, the components of that distance along the axes must each "co-vary" so as to maintain the constancy of the distance as a whole.]

6. [Einstein had already called these equations the "Galilei transformation" in his 1916 popular introduction (Einstein 1961); the term continues to be used in the slightly modified form "Galilean transformations."]

7. [Inertial systems are "favored" because, by definition, in them Newton's laws of motion hold.]

8. [Rewriting this equation slightly by moving some terms to the right-hand side gives $c^2dt^2 = dx^2 + dy^2 + dz^2$, which describes a sphere of light expanding from the origin whose radius at time dt is $c\ dt$.]

Space, Ether, and Field in Physics (1930)

Albert Einstein

Concepts of prescientific origin have always been objects of philosophical conflict. This has been especially true of the concept of space. Where does this concept come from? To what extent does it originate in experience? What experiences does it refer to? We may answer that concepts, considered logically, never originate in experience; i.e. they cannot to be derived from experience alone. And yet they are formed in our mind only with reference to what is experienced by the senses. We have to explain such fundamental concepts by pointing out the characteristic of our sense-experiences that has led to the formation of the concept.

In the case of space this connection is transparent. The concept of space is preceded by the formation of the concept of an objective world of bodies. I can recognize bodies by sensible properties without having yet grasped the bodies spatially. When the concept of body is thus formed, sense experience compels us to acknowledge relations of location among bodies, i.e. relations of mutual contact. What we interpret as spatial relations among bodies is just this. Accordingly: without the concept of body, no concept of spatial relation among bodies; and without the concept of spatial relation, no concept of space.

But how is the concept of space itself constructed? If I imagine that all bodies are, surely empty space would still remain? Or is even this concept to be made dependent on the concept of body? Yes, certainly, I reply![1] When considering the mutual relations of the location of bodies, the human mind finds it much simpler to relate the locations of all bodies to that of a single one rather than to grasp mentally the confusing complexity of the relations of every body to all others. This *one* body, which is everywhere and must be capable of being penetrated by all others in order to be in contact with all, is indeed not given to us by the senses, but we devise it as a fiction for convenience in thought. In our practical daily life, the surface of the body we call the earth plays such a part in our grasping of spatial relations among bodies that on account of it the formation of the concept of space as we have outlined it may have been rendered much easier.

This view of the concept of space as arising from the comprehension of the essential characteristic of all local relations among bodies is also confirmed by the consideration of the development of the scientific theory of space in geometry. For in the oldest geometry, which the Greeks gave us, investigation is limited solely to the local relations of idealized corporeal objects, which are called "point," "straight line," and "plane." In the concepts of "congruence" and "measurement" the reference to the local relations of corporeal objects is plainly shown. A spatial continuum, in short, "Space," is not to be found in the Euclidean geometry at all, in spite of the fact that this concept had of course been current in prescientific thought.

The extraordinary significance of the geometry of the Greeks lies in the fact that it is—as far as we know—the first successful attempt to comprehend a complex of sense-experience conceptually by means of a logically deductive system.

The spatial continuum as such was introduced into geometry by the moderns, first, that is, by Descartes, the founder of analytic geometry.

Descartes's service in introducing the spatial continuum into geometry can scarcely be overestimated. In the first place, this step rendered possible the description of geometrical figures by the aid of analysis. In the second place, it decidedly deepened the scientific character of geometry. For the straight line and the plane were henceforth no longer fundamentally privileged above other lines and surfaces, but all lines and surfaces were subject to a similar treatment. The complicated system of axioms in Euclidean geometry was replaced by a single axiom which runs as follows, in contemporary language: There are systems of coordinates in relation to which the distance ds between adjacent points P and Q is expressed in terms of the differences of the coordinates dx_1 dx_2 dx_3 by means of the Pythagorean Theorem, that is, by the formula

$$ds^2 = dx_1^2 + dx_2^2 + dx_3^2.$$

From this, that is, from the Euclidean metric, all concepts and proposition of Euclidean geometry may be deduced.

The most important point, however, is perhaps that without the introduction of the spatial continuum in the Cartesian sense, the formulation of Newton's mechanics would not have been possible. The fundamental concept utilized in Newton's theory, namely that of acceleration, has to rest on the concept of the Cartesian space of coordinates, for acceleration absolutely cannot be derived from concepts that refer only to the *relative* location of bodies or material points and their temporal change. Hence it may rightly be said that, according to Newton's theory, space plays the part of a physically real entity, as Newton was well aware, but later thinkers mostly overlooked.

Thus, from a physical standpoint, the Cartesian space of coordinates had at first two independent functions. It defined (by means of the Pythagorean Theorem or Euclidean metric) the possible locations of practically rigid bodies as well as the motion of inertia of material points. This space seemed absolute in the sense that it produced effects on things, but that nothing could produce any modification in it—the boundless, eternally unchangeable vessel of all being and all events.

Newtonian physics had its entire basis in the concepts of space, time, and mass. Every process in nature was to be understood by means of these concepts. It is true that there were added the special concepts of electric and magnetic mass, which were in many respects analogous to the original concept of mass, but seemed to be deprived of the chief characteristic of mass, namely, inertia.[2] In all other respects, however, the Newtonian view of the physically real seemed to be the only possible one.

But in the nineteenth century a reversal slowly came to pass, which gradually brought about a complete change in the conception of the physically real. When the wave-character of light, as evidenced in phenomena of interference and bending, had been demonstrated by Young and Fresnel, an all-pervading medium was needed for the light waves, which penetrate even empty space. This medium was called ether. At first it was thought of as similar to the bodies of Newtonian physics. However, its all-pervasiveness, its intangibility, and the fact of absence of friction with ponderable bodies, all imparted to it a kind of strangeness, of ghostliness, I might almost say. It seemed indubitable, yet mysterious and incomprehensible at the same time. Still worse, the mechanical properties ascribed to it were contradictory.

Newton's theoretical framework was totally shattered by the Faraday-Maxwell field theory of electromagnetic phenomena. Gradually the idea emerged that the electromagnetic fields, which are sometimes located in space devoid of all matter, cannot be consistently or satisfactorily explained as mechanical states of the ether. One became accustomed to thinking of the electromagnetic fields as fundamental realities of a nonmechanical nature. Yet they were still viewed as states of the ether. But the ether could no longer be regarded as a structure analogous to ponderable matter; the less so, since (about the turn of the century) the theory of the molecular structure of matter was more and more widely accepted, whereas the ether seemed to fill space continuously.

Although the fields themselves had become established as fundamental realities incapable of mechanical exposition, there still remained the question about the mechanical properties of their medium, the ether. This was answered by H. A. Lorentz, who showed that all electromagnetic facts compel us to regard the ether as everywhere at rest relative to Cartesian or Newtonian space. This immediately suggested the idea that the fields are states of space and that space and ether are one and the same. This was, however, not accepted. The reason was that space, the seat of Euclidean metric and of the inertia of Galileo and Newton, was regarded as absolute, that is, incapable of being affected by anything—a rigid skeleton of the world that, so to speak, is there prior to all physics and cannot be the medium of changing states. With the abandonment of a mechanical interpretation of the electromagnetic fields, these fields became independent physical realities, along with material corpuscles.

The next step in the development of the concept of space is that of the special theory of relativity. The law of the transmission of light in empty space in connection with the principle of relativity with reference to uniform movement led necessarily to the conclusion that space and time had to be combined in a unified four-dimensional continuum. For it was recognized that nothing real corresponded to the idea of simultaneous events. As Minkowski was the first to see clearly, this four-dimensional space had to be regarded as possessing a Euclidean metric which was quite analogous to the metric of the three-dimensional space of Euclidean geometry in conjuction with the use of an imaginary time-coordinate.[3] The later development, which has become known as the "general theory of relativity," is founded on the existence of a structure of space which is characterized by a Euclidean metric.

After it was realized that an absolute character cannot be ascribed to velocity or even to acceleration, it was evident that nothing real in nature corresponds to the concept of the system of inertia. It became clear that laws had to be formulated in such a way that the formulation could claim validity relative to every system of Gaussian coordinates in four-dimensional space (universal covariance of the equations expressing the laws of nature). This is the formal content of the general principle of relativity. Its heuristic force lies in the question: What are the simplest universally covariant meaningful systems of equations (covariant signifying independent of the choice of coordinates)?

The question, however, is not fruitful in so general a form. There must be added some description of the character of the structure of space. This was given by the special theory of relativity, the validity of which had to be granted for infinitely small portions of space. This means that there is a structure of space which can be expressed mathematically through a Euclidean metric for the infinitesimal environment of every point. Or: space possesses a Riemannian metric. For physical reasons, it was clear that this Riemannian metric also is the mathematical expression for the field of gravitation.

Thus, the mathematical question that corresponds to the problem of gravitation was this: What are the simplest mathematical conditions to which a Riemannian metric in four-dimensional space can conform? Thus were found the field equations of gravitation in the general theory of relativity, which have experienced the well-known verifications.

The significance of this theory for the knowledge of the nature of space may be characterized thus: Space loses its absolute character with the general theory of relativity. Up to that phase of development, space was regarded as something the inner structure of which was incapable of being affected by anything and was absolutely and completely unchangeable; hence a special ether had to be assumed as the medium of field-states located in space devoid of matter. But since the theory of relativity, the most characteristic property of space—its metrical structure—has been recognized to be changeable and capable of being affected. The condition of space attained the character of a field; geometrical space had become analogous to the electromagnetic field in this regard.[4] Thus, the separation of the concepts space and ether was abolished [*aufgehoben*], as it were, on its own accord, after the special theory of relativity had already taken from the ether the last remnant of materiality.[5]

I do not by any means find the chief significance of the general theory of relativity in the fact that it has predicted a few minute observable effects, but rather in the simplicity of its foundation and in its logical consistency. It takes away the absolute character of the space-time continuum and derives metric or geometry, inertia, and gravitation from a single structural property of four-dimensional space, namely, its Riemannian metric.

But this theoretical construction leads necessarily beyond itself. It interprets the gravitational field, it is true, as a metric structure of space. But it requires the introduction of special conceptual elements if it is to include the electromagnetic field within itself. That is to say, it does not give a logically satisfactory account of precisely those phenomena to which the whole theory of relativity owes its origin.

There is now earnest effort on the part of theoretical physicists to derive the total concept of the gravitational and electromagnetic field from a unified structure of space. What can this structure be? How can the mathematical field-laws be found to which this structure conforms? Can the material particles (electrons and protons) be conceived as regular solutions of these field-laws? These are the questions to which one is currently struggling to find answers.

As long as the questions are not satisfactorily solved, there will be a justified doubt as to whether such far-reaching deductive methods may be granted to physics at all. Only success can render a final decision on this point.

One additional remark should be made in this connection. There is available mathematically a structure of space which is a natural supplementation of the structure of space according to the Riemannian metric. I have called the investigation of this structure a "unified field theory"—thus hinting at the hope which I attach to it. Let us form a conception of this structure.

Let P und P' be any two points of the continuum, and $\overline{PQ}$ and $\overline{P'Q'}$ two line-elements proceeding from these points. The presupposition of the metrical structure means that we can speak meaningfully of the equality of the two line-elements; or, more generally, that line-elements are comparable with reference to their magnitude. The Riemannian type of metric is expressed by the presupposition that the square of the length of the line-element may be expressed by a homogeneous function of the second degree of the coordinate-differentials. On the other hand, within the framwork of Riemannian geometry a proposition about a relation of direction, e.g. parallelism, of two line-elements $\overline{PQ}$ and $\overline{P'Q'}$, has no meaning. But if there now be added the presupposition that one may speak meaningfully of a parallel-relation of line-elements, then one arrives at the formal foundation of the unified field theory.[6] For the sake of completeness there should be added the presupposition that the angle formed by two line-elements proceeding from the same point is not changed by their parallel movement.

For the mathematical expression of the field laws, we require the simplest mathematical conditions to which such a structure of space can conform. Such laws seem actually to have been discovered and they agree in fact with the empirically known laws of gravitation and electricity in first approximation. Whether these field-laws will also yield a usable theory of material particles and of motions must be determined by deeper mathematical investigations.[7]

According to the views here presented, the axiomatic foundation of physics appears as follows. The real is conceived as a four-dimensional continuum with a unified structure of a definite kind (metric and direction). The laws are differential equations, which the structure mentioned satisfies, namely, the fields which appear as gravitation and electromagnetism. The material particles are positions of high density without singularity.

We may summarize in symbolical language. Space, brought to light by the corporeal object, made a physical reality by Newton, has in the last few decades swallowed ether and time and seems about to swallow also the field and the corpuscles, so that it remains as the sole medium of reality.

Notes

[First appeared in Einstein 1930, including the original German text and an English translation by Edgar S. Brightman, reprinted here with corrections and emendations.]

1. [Notice Einstein's more emphatic statement of the dependence of space on bodies, which he states less strongly in his 1926 encyclopedia article, 164 above.]

2. [By "electric and magnetic masses" Einstein seems to mean the part of the mass-energy of particles due to their electromagnetic interactions, in contrast with to their gravitational masses (that due purely to gravitational interaction) or their inertia (the measure of their response to external force).]

3. [That is, the special relativistic expression for the line-element $ds^2 = c^2dt^2 - dx^2 - dy^2 - dz^2$, can be written $ds^2 = -d\tau^2 - dx^2 - dy^2 - dz^2$ if we define $\tau = ict = \sqrt{-1}ct$. Doing so makes the line-element take on a "quasi-Euclidean form," in that now spatial and temporal variables all enter with the same sign. For Minkowski's famous 1908 address on "Space and Time," see Lorentz et al. 1923, 75–91.]

4. [This is a sharper formulation of the new character of space *as* field than in Einstein's earlier essays in this volume.]

5. [The German term *aufgehoben* is complex and important; though one of its meaning is indeed "abolished," Hegel used *aufgehoben* to indicate the way in which a synthesis has "raised up" (another literal meaning of this term) conflicting and even contradictory states into a higher unity.]

6. [Here Einstein begins the exposition of his 1929 attempt at a unified field theory based on "distant parallelism," which Cartan will discuss in the following paper.]

7. [Unfortunately, neither this (1929) theory nor any of Einstein's later attempts met with success, see 21, note 46.]

Euclidean Geometry and Riemannian Geometry (1931)

Élie Cartan

Antiquity has left us in Euclidean geometry a finished model of deductive theory. In accord with reality as much as possible, for centuries it has satisfied the strictest exigencies of logical rigor; in the chain of its axioms, only the parallel postulate has seemed to lack the character of self-evidence that was taken as the criterion of truth. As we know, the efforts made since the end of the eighteenth century to demonstrate this postulate and thus bring geometry to the point of absolute perfection that human thought require did not succeed, but they led to a result of considerable philosophic significance. In fact, they showed, through the creation of non-Euclidean geometries that are perfectly coherent and founded on the same self-evident axioms as Euclidean geometry, that geometry cannot be deduced, as certain philosophers had wanted to believe, only from the exigencies of human reason. It is true that, though the edifice of Euclid lost its character of absolute necessity, it kept its logical solidity, provided that the parallel postulate was accorded its true character, not that of a theorem one had not succeeded to demonstrate, but that of a indemonstrable hypothesis, irreducible to axioms in the true sense.

These axioms themselves, regarded so long as self-evident truths, were submitted in their turn to close critical analysis. We discovered that some of them were simple tautologies, others were hypotheses without logical necessity, and finally, in addition to the formulated axioms were still others, not the least fundamental, implicitly admitted even without the need for their explicit enunciation being felt. Thus, we realized that the parallel postulate did not enjoy, among the axioms of geometry, the very special character that we had believed and that we might, while admitting this postulate and remaining in the general line laid out by Euclid, introduce variants into the enunciation of the "self-evident axioms" so as to give birth, next to Euclidean geometry, to other, equally legitimate geometries (non-Archimedean geometries, non-Pascalian, etc.).

Thanks to all this work of critical revision, which had its outcome, at least provisionally, with Hilbert, Euclidean geometry won the possibility of unshakeable logical bases, thus becoming a purely deductive science so perfect that the harmonious chain of relations between fundamental geometrical notions could proceed without the need to attach some concrete sense to these notions.[1] The question of the accord of Euclidean geometry with the human mind was thus completely resolved, but in a manner quite independent of the no less important question of its accord with physical reality.

II.

Exactly this last question, long before the critical work it called for was completed, preoccupied an eminent geometer, Riemann, whose creative thought left so many deep impressions in different domains of mathematics. In his inaugural dissertation "On the Hypotheses That Lie at the Foundations of Geometry" (1854), Riemann, regarding *measure* as a fundamental geometrical operation, showed the necessity of preceding all discussion of the principles of geometry with a study of the manifolds (*Mannigfaltigkeiten*) susceptible to measure, that is, those in which the length of every line can be compared quantitatively to that of every other line. He limited himself to the case of continuous manifolds with a finite number of dimensions, n, in which the position of every element or point can be defined by n numbers or *coordinates* $x_1, x_2, \ldots, x_n$, and he examined the laws or *metrics* capable a priori of furnishing the distance between two infinitely nearby points. Riemann said that experience would decide which metric suits our space. For him, that metric does not depend on the nature of *space in itself*, at least if we assume continuous space; it would probably be due to the action of binding forces exerted on bodies immersed in space: "Therefore, either the reality underlying space must form a discrete manifold, or the basis for the metric relations must be sought outside it, in binding forces acting on it." Space is thus not simply a preexisting extension, in which each body comes to occupy a quantitatively determined part. It would doubtless be excessive to push Riemann's thought so far, but it is to be doubted that he should be regarded as the precursor of Einstein, as Weyl has justly remarked.

Despite its extreme interest, let us set aside this aspect of Riemann's theory. Among all the metrics possible a priori, he regarded as the simplest that representing the distance between two infinitely nearby points as the square root of a positive definite quadratic differential form. Nowadays, we give the name of Riemannian manifolds or varieties or Riemannian spaces to manifolds that obey a metric of this kind, and Riemannian geometry is the study of the geometrical properties of these manifolds. Euclidean geometry is a Riemannian geometry because in it, following the Pythagorean Theorem, the distance between infinitely nearby points is given in rectangular coordinates by the expression $\sqrt{dx^2 + dy^2 + dz^2}$, but this Riemannian geometry is completely unique because in general it is impossible, by a suitable choice of coordinates, to reduce a quadratic differential form to a sum of squares of the differentials of the coordinates. A Riemannian metric also describes any surface immersed in our Euclidean space, when we only pay attention to the lengths of lines laid out on that surface. In this regard, different surfaces that can be deduced from one another by a deformation that preserves all these lengths constitute only a single identical Riemannian manifold in two dimensions. We can therefore say that Gauss, to whom we owe the study of surfaces from the preceding point of view, is the precursor of Riemannian geometry.

III.

At first glance, Riemannian geometries are distinguished from Euclidean geometry by a certain number of important traits.

Consider an initial difference between the Euclidean and Riemannian points of view. The global properties of space enter in from the beginning of Euclidean geometry; in affirming that two distinct straight lines cannot have two points in common, we state this whatever chain of phenomena may be associated with a straight line, as a property of all of space, not only of that portion accessible to our experience. This characteristic of Euclidean geometry is still more striking in the definition of two parallel straight lines. It is also manifest when we admit that every straight line divides the plane into two distinct parts, a hypothesis invoked more or less implicitly in the demonstration of the theorem that from a point one can erect only one perpendicular to a given straight line. Not only is Euclidean geometry a *global* geometry, but also it is accompanied with hypotheses on the nature of space considered from the point of view of analysis situs[2] (space is infinite in every sense, etc.). Riemannian geometry is, on the contrary, founded solely on the *local* properties of space, since the only source of geometrical properties is the expression of the distance of two infinitely nearby points; in this sense, Riemannian geometry harmonizes much more than Euclidean with the general tendency of modern physics and is thus compatible with every hypothesis made a priori on the form of space. It seems that, for Riemann, space is *boundless*, that is, in the language of modern topology, it is a manifold having only interior points, but which nothing obliges us to suppose *infinite* (the surface of a sphere gives us the notion of a manifold in two dimensions that is unbounded and finite). Modern analysis gives us varied examples of Riemannian spaces, whether *open* like Euclidean space, or *closed*, like the sphere, very different from each other from the qualitative point of view of analysis situs. We cannot forget, moreover, that the founder of analysis situs was Riemann himself, who showed its fundamental importance in the theory of functions, a discipline that proves more and more fruitful among the diverse branches of mathematics.

We have a quite remarkable illustration of this opposition between the local character of Riemannian geometry and the global character of Euclidean geometry precisely when we study Euclidean geometry *as* Riemannian geometry. If for the distance between two infinitely nearby points we accept the same analytic law as in Euclidean geometry, we will recover Euclidean space on the condition of assuming that it is infinite and simply connected,* but if we set aside these hypotheses, we find other spaces that are indistinguishable

*This means that every closed curve in space can be reduced to a point by continuous deformation.

from Euclidean space with respect to their geometrical properties, as long as we remain in a finite region that is sufficiently small, but which differ from it profoundly when considered as a whole. The simplest example is furnished by the surface of a cylinder of revolution, considered as a Riemannian space of two dimensions; Clifford has given another example of a space that is *locally Euclidean* two dimensions and identical, from the point of view of analysis, to the surface of a torus.[3] There are others still more paradoxical. Because our experience is necessarily limited, it seems we shall never be able to affirm that we do not live in one of these paradoxical spaces.

From the *local* or *differential* character of Riemannian geometry, opposed to the *global* character of Euclidean geometry, flows a fundamental difference in the methods proper to the study of these two geometries. In place of the synthetic methods of the ancients must be substituted (at least at first) purely analytic methods, because the origin of geometrical properties is in an analytic expression given a priori. Euclidean geometry itself also uses analytic methods relying on the use of coordinates, but these coordinates (Cartesian, rectangular, polar, etc.) have a precise, quantitative geometric significance, which is why they can be introduced only after the geometry is founded by its own methods. In Riemannian geometry, on the contrary, coordinates, introduced from the beginning, serve simply to relate empirically the different points of space, and geometry has precisely the object of extricating the geometric properties that are independent of this arbitrary choice of coordinates. Problems of this nature are not absolutely new. Gauss, in his *General Investigations of Curved Surfaces*, had given just such a model of development of Riemannian geometry in two dimensions. On the other hand, Lamé, several years (it is true) after Riemann's dissertation, had introduced coordinates for *Euclidean* geometry in three dimensions not having as great a degree of generality as Riemann's, but, in this somewhat more restricted case, he also found Riemannian geometry.

The necessity of using systems of arbitrary coordinates exerted a profound influence on the later development of mathematics and physics. It led to the admirable creation of an absolute differential calculus by Ricci and Levi-Civita, which was the instrument that helped to elaborate general relativity. Nonetheless, it is fitting that we not forget that the degree of certainty of geometrical or physical theory does not reside, as some have seemed to believe, in the generality of the system of coordinates in which this theory finds its analytic expression.

Let us draw attention to a final distinctive characteristic of Euclidean space and Riemannian space. The former is *homogenous*, the latter in general not. A body immersed in Euclidean space has geometric properties that are conserved when one *displaces* the body in space; the notion of *displacement* plays a fundamental role in Euclidean geometry, closely tied to the notion of geometric *equality* of which it is basically only a particular aspect. The axiom according to which two figures equal to a third are equal to each other is only a way of stating the property of displacements to form a *group*. The notion of group thus underlies the entire edifice of Euclidean geometry and we know how Lie and Hilbert used this notion to reconstitute geometry.[4] One can say that Euclidean geometry has the object of studying the properties of figures invariant under the operations of a certain group. In Riemannian geometry, there is nothing like this for, as Riemann puts it, a body has no existence apart from the position it occupies. Moreover, Riemann himself determined all the *homogenous* Riemannian spaces, or rather those in which bodies have the same degree of mobility as in Euclidean space; he recovered in this way the non-Euclidean space of Lobachevsky and added to it spherical space, boundless but finite.

Riemann's conception of a nonhomogenous space has great philosophical importance; affirming, along with Einstein, this conception as a result of experience is to deny the principle of causality, as Painlevé understands it, according to which the same causes acting in two distinct regions of space and at two different instants in time must produce the same effects. To tell the truth, this removes all meaning from the very statement of this principle.

IV.

However deep may be the differences between the two geometries, Euclidean and Riemannian, they have common characteristics that the very elaboration of Riemannian geometry has shown to be more and more significant. In order better to compare them to each other, it is necessary to consider them in the domain common to both, that of differential geometry. In this regard, we can conceive a geometer, placed in a Riemannian space, but having a Euclidean turn of mind, who seeks to account to himself for the geometrical facts that can be observed. He will be somewhat in the situation of a contemporary physicist whose mind, formed by a long Euclidean heritage, is obliged by experience to modify, following Einstein, the conception he made of spatio-temporal phenomena.

Our Euclidean geometer, if he remains immobile and considers only what happens in his immediate neighborhood, has no reason to modify his conceptions; in fact, the analytic expression that gives the distance from point O where he is standing to an infinitely close point M is exactly the same as in a Euclidean space that is related to a system of Cartesian coordinates. The geometer can then believe himself in the middle of a Euclidean space: all the laws that regulate the lengths of very small vectors from the origin O, the angles made by these vectors, the areas of the surface elements issuing from O are exactly the same in the Riemannian space in which it is really immersed as in the Euclidean space in which he believes himself to be immersed. The little piece of Riemannian space that surrounds him is a piece of Euclidean space and we can say, in this sense, that Riemannian space as a whole is formed of an infinity of infinitesimally small pieces of Euclidean space.

Our geometer can go further; moving along the arc length of a curve OA, he can calculate more and more closely this arc length; comparing the lengths of different curved arcs leading from O and A, he can determine the shortest, which will be in this case a segment of a straight line or, as Riemann puts it, a segment of a geodesic. He can also construct a system of rectangular coordinates. He will choose three rectangular directions issuing from O; the geodesic joining O to any point whatever M of the space will be defined, as in Euclidean geometry, by the cosines α, β, γ of the three angles made at O with the three directions of the coordinates.[5] If s is the length of OM, we will quite naturally take as the coordinates of M the three quantities $\alpha s, \beta s, \gamma s$, just as in Euclidean geometry. This way of forming an intrinsic system of coordinates from the origin O (*normal* coordinates) was indicated by Riemann himself. With the aid of these coordinates, our geometer with a Euclidean turn of mind can make a sort of *map* of the Riemannian space on the ideal Euclidean space in which he believes he lives. This map reproduces faithfully the distances, calculated following the shortest lines, that separate point O from any point M, the angles between these lines, etc., but it alters the other distances. For example, it could be that the distance between two points M and N nearby O are shorter, or longer, on the map than in reality. If points M and N remain in the same *plane element* that issues from O, the alteration of their distance is proportional to the area, measured on the map, of the triangle OMN; the factor of proportionality measures the *Riemannian curvature* of the space in the direction of the plane element considered and is positive if the map enlarges the distances, negative in the contrary case. All these laws were determined by Riemann.[6]

Instead of proceeding like Riemann, our geometer would be able, as he walked, to pursue his experiences without losing his Euclidean illusions. If in a Euclidean space, two nearby observers each adopt an orthogonal system of axes,[7] geometry will furnish a way to refer these two systems of axes to each other and thus compare their respective orientations, for example to decide if some direction with respect to the origin of the first axis is parallel to some direction with respect to the second. Levi-Civita made immense progress with Riemannian geometry when he extended to every Riemannian space the possibility of comparison by means of his theory of parallel transport. In this way, two small pieces of Euclidean space that surround two nearby observers can be integrated into one and the same Euclidean space and this integration can be extended further and further if one has a continuous linear series of observers. To put it otherwise, our geometer walking along line C and only measuring along this line and in its immediate neighborhood, can legitimately believe himself immersed in a Euclidean space. If he attaches to every point M of the line an orthogonal system of axes to which the measurements made from M refer, he can imagine in his ideal Euclidean space a faithful map of line C and all the axes attached to it; he will thus know the curvature and the torsion of this line, etc.

In conclusion, *all the operations of Euclidean differential geometry are transportable, just as they are, to Riemannian geometry*, at least as long as one moves along a line. How will our geometer, whose experiences have not yet troubled his Euclidean prejudices, ever be forewarned of his error? By this: if, starting from the axes attached to a point O, he makes a map of two curved arcs C and C' going from O to the same point A, these two arcs each being provided with axes attached to their different points, *he will not find* in the two cases the same position for point A and the axes attached to it: it will be necessary, in order to pass from one position to the other, to make a certain Euclidean displacement that he decompose into a translation and a rotation. These two operations define what Cartan calls the *torsion* and the *curvature* associated with the closed contour determined by the two arcs C and C' and running between one and the other.[8] If this contour is infinitely small, the torsion vanishes (or rather is infinitely small in proportion to the area bounded by the contour) and the curvature is defined analytically by the same quantities that Riemann had attached to the plane element in which the contour lies, but these quantities now have a *Euclidean* interpretation, since they represent a rotation undergone by an orthogonal system of axes in a Euclidean space.

V.

All the preceding permits us to say that Riemannian geometry is a *nonholonomic* Euclidean geometry.[9] But there is more. There exists in a Riemannian space an infinity of laws, different from those of Levi-Civita, permitting us to relate two infinitely nearby orthogonal systems of axes to each other, so that from the point of view in which we finally are placed, the given quadratic differential form defining the square of the distance between two infinitely nearby points does not determine completely all of the geometric structure of the space: we must add to it the law of interrelation of two infinitely nearby axes. Depending on which law we choose, space can take in each plane direction a different curvature and torsion; the law of Levi-Civita is the only one that does not include any torsion in space. The latest theories of Einstein use just such a law that does not include any curvature.

But what is even more important from the preceding point of view is that we can create beside every geometry founded on a group—such as projective geometry, for example—an infinity of other nonholonomic geometries that play with respect to projective geometry the role that Riemannian geometry plays with respect to Euclidean geometry. The possibility of this generalization rests simply on the possibility of relating one with respect to the other— in projective geometry, two nearby pieces of space—and this possibility has for its origin the homogeneity of projective space, that is, the existence of a group at the foundations of projective geometry. In the last analysis, the synthesis of the apparently completely divergent master ideas of Riemann and Lie-Klein thus is at the origin of the entire development of modern differential geometry.

For now, all these new geometries interest scarcely anyone beside mathematicians, but in a period when physicists are on the look-out for the most abstract mathematical creations, nothing says that one day they will not furnish a theory able to tie together the mass of observed new phenomena. As Riemann says himself, experience seems, in the domain of things that are not extremely small on our scale, to justify our belief in the Euclidean structure (holonomic or nonholonomic, we may add) of space, but nothing authorizes us to affirm that it is thus in the domain of the infinitely small.

Moreover, there are grounds for deepening Riemannian geometry, properly called, in another direction. Its entire development was based on an unwarranted hypothesis, namely that the functions that enter the analytic expression of the distance between two infinitely nearby points allow derivatives of several orders. It is very probable that the general appearance of Riemannian geometry will change completely if one allows only the continuity of these functions. In this lie many very difficult problems that can be attacked only by using all the resources of modern analysis.

Notes

[From Cartan 1931, which does not appear in his collected works, Cartan 1955.]

1. [See Hilbert 1959.]
2. [Here Cartan uses the older term analysis situs for what (following Poincaré) is now called topology, a term introduced in 1847 by J. B. Listing, which Cartan also uses in this essay.]
3. [See Klein's figure and discussion above, 113.]
4. [Cartan, like Poincaré, does not identify this approach as the "Erlangen Program" and connects it with Lie but not Klein.]
5. [Now these angles are generally called "direction cosines."]
6. [Riemann generalized these results from Gauss's initial discoveries of their analogues for curved two-dimensional planes, now called the Gauss-Bonnet Theorem; see Gauss 2005, vi, 30, 48.]
7. [Cartan uses the expression "trihedral reference tetrahedron" (*trièdre trirectangle de référence)* to denote what is generally called an orthogonal system of axes, which I have used to render this phrase; elsewhere (Cartan 1931, 15) he does use the more transparent term "système de reference," a more obvious equivalent for "system of reference." I thank Jean-Marc Lévy-Leblond and Olivier Darrigol for their helpful advice about translating these terms.]
8. [Here Cartan refers to his own work; see Cartan 1983, 1986.]
9. [Cartan means here by "nonholonomic" what he just described at the end of section IV: in Riemannian geometry, if we transport a vector at O parallel to itself to A via two different paths C and C', in general we will not get the same resulting vector. In the language of modern differential geometry, we can assign to each point of a Riemannian manifold a Euclidean space (a "vector bundle," such as the "tangent bundle" at that point) and then the "connection" allows us to string together these different Euclidean spaces so that we can map how a vector from a bundle at one point goes into a vector from the bundle at another as we travel along a certain path. This will, in general, depend on the path taken, so that if we take a closed path, which returns to the same point, the initial vector may not be mapped onto itself identically, which Cartan calls "nonholonomic." I thank Barry Mazur for helping clarify and illustrate this term.]

The Problem of Space, Ether, and the Field in Physics (1934)

Albert Einstein

Scientific thought is a development of pre-scientific thought. As the concept of space already played a fundamental role in the latter, we must begin with the concept of space in pre-scientific thought. There are two ways of regarding concepts, both of which are necessary to understanding. The first is that of logical analysis. It answers the question: How do concepts and judgments depend on each other? In answering it, we are on comparatively safe ground. It is the certainty by which we are so much impressed in mathematics. But this certainty is purchased at the price of emptiness of content. Concepts can only acquire content when they are connected, however indirectly, with sensible experience. But no logical investigation can reveal this connection; it can only be experienced. And yet it is this connection that determines the cognitive value of systems of concepts.

For example, suppose an archaeologist belonging to a later culture finds a textbook of Euclidean geometry without diagrams. He will discover how the words "point," "straight line," "plane" are used in the propositions. He will also see how the latter are deduced from each other. He will even be able to frame new propositions according to the known rules. But the framing of these propositions will remain an empty word-game for him, as long as "point," "straight line," "plane," etc., convey nothing to him. Only when they do convey something will geometry possess any real content for him. The same will be true of analytical mechanics, and indeed of any exposition of a logically deductive science.

What does it mean that "straight line," "point," "intersection," etc. convey something? It means that one can point to the sensible experiences to which those words refer. This extra-logical problem is the essential problem, which the archeologist will only be able to solve by intuition, by examining his experience and seeing if anything he can discover corresponds to those primary terms of the theory and the axioms laid down for them. Only in this way can the question of the nature of a conceptually presented entity be reasonably raised.

Regarding the problem of the nature of the pre-scientific concepts in our thinking, we are almost in the position of that archaeologist. We have, so to speak, forgotten what features in the world of experience caused us to frame those concepts, and we have great difficulty in representing the world of experience to ourselves without the spectacles of the old-established conceptual interpretation. There is the further difficulty that our language has to work with words which are inseparably connected with those original concepts. These are the obstacles that confront us too when we try to describe the nature of the pre-scientific concept of space.

One remark about concepts in general, before we turn to the problem of space: concepts have reference to sensible experience, but they are never, in a logical sense, deducible from them. For this reason I have never been able to comprehend the problem of the a priori as posed by Kant. With regard to any problem concerning the nature of concepts, the only possible procedure is to seek out those characteristics in the complex of sense experiences to which the concepts refer.

Now as regards the concept of space: this seems to be preceded by the concept of the solid object. The characteristics of the complexes and sense impressions that are probably responsible for that concept has often been described. The correspondence between certain visual and tactile impressions, the fact that they can be continuously followed out through time, and that the impressions can be repeated at any moment (taste, sight), are some of those characteristics. Once the concept of the solid object is formed in connection with the experiences just mentioned—which concept by no means presupposes that of space or spatial relation—the desire to get an intellectual grasp of the relations of such solid bodies to each other is bound to give rise to concepts which correspond to their spatial relations. Two solid objects may touch one another or be distant from one another. In the latter case, a third body can be inserted between them without altering them in any way, in the former not. These spatial relations are obviously real in the same sense as the bodies themselves. If two bodies are of equal value for the filling of one such interval, they will also prove of equal value for the filling of other intervals. The interval is thus shown to be independent of the selection of any special body to fill it; the same is universally true of spatial relations. It is evident that this independence, which is a main prerequise for the usefulness of framing purely geometrical concepts, is not necessary a priori. In my opinion, this concept of the interval, detached as it is from the selection of any special body to occupy it, is the starting point of the whole concept of space.

Considered, then, from the point of view of sense experience, the development of the concept of space seems, after these brief indications, to conform to the following schema: solid body; spatial relations of solid bodies; interval; space. Looked at in this way, space appears as something real in the same sense as solid bodies.

It is clear that the concept of space as a real thing already existed in the extra-scientific conceptual world. Euclid's mathematics, however, knew nothing of this concept as such; it made do solely with the concepts of the [solid] object and the spatial relations between objects. The point, the plane, the straight line, length, are solid objects idealized. All spatial relations are reduced to those of contact (the intersection of straight lines and planes, points lying on straight lines, etc.). Space as a continuum does not figure in the conceptual system at all.

This concept was first introduced by Descartes, when he described the point-in-space by its coordinates. Here for the first time geometrical figures appear, as it were, as parts of infinite space, which is conceived as a three-dimensional continuum.

The great superiority of the Cartesian treatment of space is by no means confined to the fact that it applies analysis to the purposes of geometry. The main point seems rather to be this:—The geometry of the Greeks prefers certain figures (the straight line, the plane) in geometrical descriptions; other figures (e.g., the ellipse) are only accessible to it because it constructs or defines them with the help of the point, the straight line, and the plane. In the Cartesian treatment, on the other hand, all surfaces are, in principle, equally represented, without any arbitrary preference for linear figures in building up geometry.

Insofar as geometry is conceived as the science of laws governing the mutual positions of practically rigid bodies, it is to be regarded as the oldest branch of physics. This science was able, as I have already observed, to get by without the concept of space as such, the ideal corporeal forms—point, straight line, plane, length—being sufficient for its needs. On the other hand, space as a whole, as conceived by Descartes, was absolutely necessary to Newtonian physics. For dynamics cannot manage with the concepts of the mass point and the temporally variable distance between mass points alone. In Newton's equations of motion, the concept of acceleration plays a fundamental part, which cannot be defined by the temporally variable intervals between points alone. Newton's acceleration is only thinkable or definable in relation to space as a whole. Thus to the geometrical reality of the concept of space, a new inertia-determining function of space was added. When Newton described space as absolute, he no doubt meant this reality of space, which made it necessary for him to attribute to it a quite definite state of motion, which yet did not appear to be fully determined by the phenomena of mechanics. This space was conceived as absolute in another sense also; its inertia-determining effect was conceived as autonomous, i.e., incapable of being influenced by any physical circumstance whatever; it affected masses, but nothing affected it.

And yet in the minds of physicists, space remained until the most recent time simply the passive container of all events, without taking part in physical occurrences. Conceptual development only began to take a new turn with the wave theory of light and the Faraday-Maxwell theory of the electromagnetic field. It became clear that there existed in free space conditions that propagated themselves in waves, as well as localized fields that were able to exert force on electrical masses or magnetic poles brought to the spot. Because it would have seemed utterly absurd to the physicists of the nineteenth century to attribute physical functions or states to space itself, they invented a medium pervading the whole of space, on the model of ponderable matter—the ether, which was supposed to act as a medium for electromagnetic phenomena, and hence for those of light also. The states of this medium, imagined as constituting the electromagnetic fields, were at first thought of mechanically, on the model of the elastic deformations of rigid bodies.[1] But this mechanical theory of the ether was never quite successful and gradually one got used to abandoning a clearer interpretation of the nature of the etheric fields. The ether thus became a kind of matter whose only function was to act as a medium (*Träger*) for electrical fields, which were by their very nature not further analyzable. The picture was, then, as follows: Space is filled by the ether, in which the material corpuscles or atoms of ponderable matter swim; the atomic structure of the latter had been securely established by the turn of the century.

Since the reciprocal action of bodies was supposed to be accomplished through fields, there had also to be a gravitational field in the ether, whose field-law, however, assumed no clear form at that time. The ether was only accepted as the seat of all operations of force that make themselves effective across space. Since it had been realized that electrical masses in motion produce a magnetic field, whose energy acted as a model for inertia, inertia also appeared as a field-action localized in the ether.

The mechanical properties of the ether were at first a mystery. Then came H. A. Lorentz's great discovery. All the phenomena of electro-magnetism then known could be explained on the basis of two assumptions: that the ether is firmly fixed in space—that is to say, unable to move at all, and that electricity is firmly lodged in the mobile elementary particles. Today his discovery may be expressed as follows: Physical space and the ether are only different terms for the same thing; fields are physical states of space. For if no particular state of motion belongs to the ether, there does not seem to be any ground for introducing it as an entity of a special sort alongside space. But physicists were still far removed from such a way of thinking; space was still, for them, a rigid, homogeneous something, susceptible of no change or states. Only the genius of Riemann, solitary and uncomprehended, by the middle of the last century already broke through to a new conception of space, in which space was deprived of its rigidity and in which its power to take part in physical events was recognized as possible. This intellectual achievement commands our admiration all the more for having preceded the Faraday-Maxwell field theory of electricity. Then came the special theory of relativity with its recognition of the physical equivalence of all inertial systems. The inseparableness of time and space emerged in connection with electrodynamics, or the law of the propagation of light. Hitherto it had been tacitly assumed that the four-dimensional continuum of events could be split up into time and space in an objective manner—that is, within the world of events, the "now" is assigned an absolute meaning. With the discovery of the relativity of simultaneity, space and time were merged in a single continuum in the same way as the three dimensions of space had been before. Physical space was increased to a four-dimensional space, which also included the dimension of time. The four-dimensional space of the special theory of relativity is just as rigid and absolute as Newton's space.

The theory of relativity is a fine example of the fundamental character of the modern development of theoretical science. The hypotheses with which it starts become steadily more abstract and remote from experience. On the other hand, it gets nearer to the grand aim of science, which is to encompass the greatest possible extent of empirical content by logical deduct ion from the smallest possible number of hypotheses or axioms. Meanwhile, the train of thought leading from the axioms to the empirical content or verifiable consequences gets steadily longer and more subtle. The theoretical scientist is compelled in an increasing degree to be guided by purely mathematical, formal considerations in his search for a theory, because the physical experience of the experimenter cannot lift him into the regions of highest abstraction. Instead of the predominantly inductive methods appropriate to the youth of science, now tentative deduction takes place. Such a theoretical structure needs to be very thoroughly elaborated before it can lead to conclusions that can be compared with experience. Here too the observed fact is undoubtedly the supreme arbiter, but it cannot pronounce sentence until the wide chasm separating the axioms from the verifiable consequences has been bridged by much intense, hard thinking. The theorist has to set about this gigantic task in the clear consciousness that his efforts may only be destined to deal the death blow to his theory. The theorist who undertakes such a labor should not be reprimanded as "fanciful"; on the contrary, he should be encouraged to give free reign to his fancy, for there is no other way to the goal. His is no idle daydreaming, but a search for the logically simplest possibilities and their consequences. This *captatio benevolentiae*[2] was needed in order to make the hearer or reader more willing to follow the ensuing train of ideas with attention; it is the line of thought which has led from the special to the general theory of relativity and thence to its latest offshoot, the unified field theory. In this exposition, however, the use of mathematical symbols cannot be entirely avoided.

We start with the special theory of relativity. This theory is still based directly on an empirical law, that of the constant velocity of light. Let P be a point in empty space, P' an infinitely close point at distance $d\sigma$. Let a flash of light be emitted from P at a time t and reach P' at a time $t + dt$. Then

$$d\sigma^2 = c^2 dt^2.$$

If dx_1, dx_2, dx_3 are the orthogonal projections of $d\sigma$ and the imaginary time coordinate $\sqrt{1}ct = x_4$ is introduced, then the abovementioned law of the constancy of the propagation of light takes the form

$$ds^2 = dx_1^2 + dx_2^2 + dx_3^2 + dx_4^2 = 0.$$

Because this formula expresses a real situation, we may attribute a real meaning to the quantity ds, even supposing the neighboring points of the four-dimensional continuum are selected in such a way that the ds belonging to them does not vanish. This is more or less expressed by saying that the four-dimensional space (with imaginary time coordinates) of the special theory of relativity possesses a Euclidean metric.[3]

The fact that such a metric is called Euclidean is connected with the following. The position of such a metric in a three-dimensional continuum is fully equivalent to the positulation of the axioms of Euclidean geometry. The defining equation of the metric is thus nothing but the Pythagorean Theorem applied to the differentials of the coordinates.

Such alteration of the coordinates (by a transformation) is permitted in the special theory of relativity, since in the new co-ordinates too the magnitude ds^2 (fundamental invariant) is expressed in the new differentials of the coordinates by the sum of the squares. Such transformations are called Lorentz transformations.

The heuristic method of the special theory of relativity is characterized by the following principle: Only those equations are admissible as an expression of natural laws that do not change their form when the coordinates are changed by means of a Lorentz transformation (covariance of equations in relation to Lorentz transformations).

This method led to the discovery of the necessary connection between impulse and energy, the strength of an electric and magnetic field, electrostatic and electrodynamic forces, inertial mass and energy, and the number of independent concepts and fundamental equations of physics was thereby reduced.

This method pointed beyond itself: Is it true that the equations that express natural laws are covariant in relation to Lorentz transformations only and not in relation to other transformation? Well, formulated in that way the question really means nothing, since every system of equations can be expressed in general coordinates. One must ask: Are not the laws of nature so constituted that they receive no essential simplification through the choice of any one particular set of coordinates?

We will only mention in passing that our empirical principle of the equality of inertial and heavy masses prompts us to answer this question in the affirmative. If we elevate [*erhebt*] the equivalence of all coordinate systems into a principle for the formulation of natural laws , we arrive at the general theory of relativity, provided we adhere to the law of the constant velocity of light or to the hypothesis of the objective significance of the Euclidean metric, at least for infinitely small portions of four-dimensional space.

This means that for finite regions of space the (physically significant) existence of a general Riemannian metric is postulated according to the formula

$$ds^2 = \sum_{\mu\nu} g_{\mu\nu} dx^\mu dx^\nu,$$

where the summation is to be extended to all index combinations from 11 to 44.

The structure of such a space differs absolutely radically in one respect from that of a Euclidean space. The coefficients $g_{\mu\nu}$ are for the time being any functions whatever of the coordinates x_1 to x_4, and the structure of the space is not really determined until these functions $g_{\mu\nu}$ are really known. One can also say: the structure of such a space as such is completely undetermined. It is only determined more closely by specifying laws that the metrical field of the $g_{\mu\nu}$ satisfy. On physical grounds, this gave rise to the conviction that the metrical field was at the same time the gravitational field.

Because the gravitational field is determined by the configuration of masses and changes with it, the geometric structure of this space is also dependent on physical factors. Thus according to this theory space is—exactly as Riemann guessed—no longer absolute; rather, its structure depends on physical influences. (Physical) geometry is no longer an isolated self-contained science like the geometry of Euclid.

The problem of gravitation was thus reduced to a mathematical problem: it was required to find the simplest fundamental equations that are covariant in relation to any transformation of coordinates whatever. This is a well-defined problem that could at least be solved.

I will not speak here of experimental confirmation of this theory, but explain at once why the theory could not rest permanently satisfied with this success. Gravitation had indeed been traced to the structure of space, but besides the gravitational field there is also the electromagnetic field. This had, to begin with, to be introduced into the theory as an entity independent of gravitation. Additional terms that took account of the existence of the electromagnetic field had to be included in the fundamental equations for the field. But the idea that there were two structures of space independent of each other, the metric gravitational and the electromagnetic, was intolerable to the theoretical mind. One is driven to the belief that both sorts of field must correspond to a unified structure of space.

Notes

[Originally appeared in German in 1934 as "Das Raum-, Äether- und Feld-Problem der Physik," in Einstein 1953, 181–193, then translated into English in Einstein 1954, by Alan Harris, revised by Sonja Bargmann (Einstein 1982, 276–285), here revised and corrected. In Einstein 1953, 266, the editor remarks that this essay's original version was the 1930 essay reproduced above, 173–178.]

1. [Here Einstein may be thinking particularly of the mechanical models that Maxwell used in his initial presentations of his field equations; see Everitt 1975, 93–110.]

2. [This Latin term from ancient rhetorical theory denotes the "capturing of the (audience's) goodwill," the speaker's plea for his hearers' sympathy.]

3. [Note that in this paper (though not in the earlier papers in this anthology) Einstein will use $d\sigma$ to denote the line-element using normal space-time coordinates and ds the line-element when the imaginary time substitution has been applied, $x_4 = \sqrt{-1}ct$; in later terminology, ds would here denote the "Euclidean" line-element. This shift in terminology may show the way in which Einstein, in his later thinking, wanted to privilege the Euclidean (imaginary time) formulation.]

References

Note that 2:111 refers to volume 2, page 111; the abbreviation *ECP* refers to Einstein's *Collected Papers.*

Abbott, Edwin A. *Flatland: A Romance of Many Dimensions*, introduction by Banesh Hoffmann. Dover, 1992.

Adler, Ronald. Maurice Bazin, and Menahem Schiffer. *Introduction to General Relativity*, second edition. New York: McGraw-Hill, 1975.

Archibald, Thomas. "Riemann and the Theory of Electrical Phenomena: Nobili's Rings," *Centaurus* **34**, 247–271 (1991).

Aristotle. *The Complete Works of Aristotle*, edited by Jonathan Barnes. Princeton: Princeton University Press, 1984.

Ashtekar, Abhay. "Quantum Geometry and its Ramifications," in Ashtekar 2005b, 350–381.

———, (editor). *100 Years of Relativity: Space-Time Structures: Einstein and Beyond.* Singapore: World Scientific, 2005b.

Balibar, F. "Geometrie und Erfahrung," in Boi et al. 1992, 91–97.

Beltrami, Eugenio. *Saggio di interpretazione della Geometria Non-Euclidea.* Naples, 1868a.

———. "Theoria fondamentale degli spazii di Curvatura costante," *Annali di Matematica*, ser. II, **2**, 232–255 (1868b).

———. *La découverte de la géométrie non euclidienne sur la pseudosphère: Les lètttres de Eugenio Beltrami a Jules Hoüel (1868–1881).* Paris: Blanchard, 1998.

Berestetskĭi, V. B., E. M. Lifshitz, and L. P. Pitaevskĭi, *Relativistic Quantum Theory*, translated by J. B. Sykes and J. S. Bell. Oxford: Pergamon, 1971.

Berger, Marcel. *Riemannian Geometry in the Second Half of the Twentieth Century.* Providence, Rhode Island: American Mathematical Society, 2000.

Bernardo, Antonio. "La geometria di Bernhard Riemann," *Cultura e scuola* **31**, 252–269 (1992).

Blackett, Donald W. *Elementary Topology: A Combinatorial and Algebraic Approach.* New York: Academic Press, 1967.

Boi, Luciano. "L'Espace: Concept Abtait et/ou physique; la Géometrie entre Formalisation Mathématique et Etude de la Nature," in Boi, Flament, and Salanskis 1992, 65–90.

————. "Die Beziehungen zwischen Raum, Kontinuum und Materie im Denken Riemanns: die Äethervorstellung und die Einheit der Physik; Das Entstehen einer neuen Naturphilosophie," *Philosophia Naturalis* **31**, 171–216 (1994).

————. "Le concept de variété et la nouvelle géometrie de l'espace dans la pensée de Bernhard Riemann: L'emergence d'une nouvelle vision des mathématiques et de ses rapports avec les sciences fondamentales," *Archives internationales d'histoire des sciences* **45**, 82–128 (1995).

————. "Die neuen geometrischen Auffassungen von Riemann bis Poincaré," *Organon* **25**, 13–38 (1995b).

Boi, Luciano, D. Flament and J.-M. Salanskis (editors). *1830–1930: A Century of Geometry: Epistemology, History and Mathematics.* Berlin: Springer-Verlag, 1992.

Bonola, Roberto. *Non-Euclidean Geometry: A Critical and Historical Study of its Development.* Dover, 1955.

Bottazzini, Umberto. "Riemanns Einfluß auf E. Betti und F. Casorati," *Archive for History of Exact Science* **18**, 27–37 (1977).

————. "Geometry and 'metaphysics of space' in Gauss and Riemann," in Poggi and Bossi 1994, 15–29.

————. "Riemann 'filosofo naturale'," *Rivista di filosofia* **87**, 129–141 (1996).

————. "Ricci and Levi-Civita: From Differential Invariants to General Relativity," in Gray 1999a, 241–259.

Bottazzini, Umberto and Rossana Tazzioli, "*Naturphilosophie* and its Role in Riemann's Mathematics," *Revue d'histoire des mathématiques* **1**, 3–38 (1995).

Boyer, Carl B. *A History of Mathematics*, second edition, revised by Uta C. Merzbach. New York: Wiley, 1991.

Brasch, Frederick E. "Einstein's Appreciation of Simon Newcomb," *Science* **69**, 248–249 (1929).

Buchwald, Jed Z. "The Quantitative Ether in the First Half of the Nineteenth Century," in Cantor and Hodge 1981, 215–238.

————. *From Maxwell to Microphysics.* Chicago: University of Chicago Press, 1985.

Cahan, David (editor). *Hermann von Helmholtz and the Foundations of Nineteenth-Century Science.* Berkeley: University of California Press, 1993.

Callender, Craig and Nick Huggett (editors). *Physics Meets Philosophy at the Planck Scale: Contemporary Theories in Quantum Gravity.* Cambridge: Cambridge University Press, 2001.

Caneva, Kenneth L. "Physics and *Naturphilosophie*: A Reconnaissance," *History of Science* **35**, 35–107 (1997).

Cantor, G. and M. Hodge (editors). *Concepts of the Ether.* Cambridge: Cambridge University Press, 1981.

Carrier, Martin. "Geometric Facts and Geometric Theory: Helmholtz and 20th-Century Philosophy of Physical Geometry," in Krüger 1994, 276–291.

———. *The Completeness of Scientific Theories. On the Derivation of Empirical Indicators within a Theoretical Framework: The Case of Physical Geometry.* Dordrecht: Kluwer, 1994b.

Cartan, Élie. *On Manifolds with an Affine Connection and the Theory of General Relativity*, translated by A. Magnon and A. Ashtekar. Naples: Bibliopolis, 1986.

———. *Geometry of Riemann Spaces*, edited by R. Hermann, translated by J. Glazebrook. Brookline, Mass.: Mathematical Sciences Press, 1983.

———. "Géométrie Euclidienne et géométrie Riemannienne," *Scientia* 49, 393–402 (1931).

———. "Le parallélisme absolu et la théorie unitaire du champ," *Revue de Métaphysique et de Morale* **38**, 13–28 (1931), also in Cartan 1952–1955, 3(2):1167–1185.

———. *Œuvres complètes.* Paris: Gauthier-Villars, 1952– 1955

Cartan, Élie and Albert Einstein. *Élie Cartan—Albert Einstein: Letters on Absolute Parallelism, 1929–1932.* R. Debever (editor) Princeton: Princeton University Press, 1979.

Castellana, Mario. "Enriques intreprete di Riemann: geometrie e filosofia" in Cimino, Sanzo, and Sava 1991, 249–272.

Chikara, Sasaki et al. (editors). *Intersection of History and Mathematics.* Boston: Birkhäuser, 1994.

Cimino, Guido,Ubaldo Sanzo, and Gabriella Sava (editors). *Il nucleo filosofico della scienza: Atti del Seminario di storia e filosofia della scienza dell'Università di Lecce (1987–1990).* Lecce: Congedo Editore, 1991.

Clifford, William Kingdon. *Seeing and Thinking.* London: Macmillan 1879.

———. *Mathematical Papers*, edited by Robert Tucker with an introduction by H. J. Stephen Smith. Bronx, NY: Chelsea, 1968 [reprint of 1882 edition].

———. *Lectures and Essays*, edited by Leslie Stephen and Frederick Pollock. London: Macmillan 1886.

Cooke, Roger. *The History of Mathematics: A Brief Course*, second edition. New York: Wiley-Interscience, 2005.

Coolidge, Julian Lowell. *A History of Geometrical Methods.* Dover, 1963.

Corry, Leo. "Hilbert and Physics (1900–1915)," in Gray 1999a, 145–188.

Daniels, Norman. "Lobachvsky: Some Anticipations of Later Views on the Relation between Geometry and Physics," *Isis* 66, 75–87 (1975).

Daston, Lorraine. "The Physicalist Tradition in Nineteenth Century French Geometry," *Studies in the History and Philosophy of Science* **17**, 269–295 (1986).

Dedekind, Richard. "Analytische Untersuchungen zu Bernhard Riemann's Abhandlungen Über die Hypothesesn, welche der Geometrie zu Grunde liegen" [including a French translation of Dedekind's paper and a German letter dated 1875 from Elisa Riemann to Dedekind], *Revue d'histoire des sciences* **43**, 237–294 (1990).

Demidov, Sergei S., M. Folkerts, D. E. Rowe, C. J. Scriba (editors). *Amphora: Festschrift für Hans Wussing.* Basel: Birkhäuser, 1992.

Derbyshire, John. *Prime Obsession: Bernhard Riemann and the Greatest Unsolved Problem in Mathematics.* Washington, D. C.: Joseph Henry Press, 2003.

Dingler, Hugo. "H. Helmholtz und die Grundlagen der Geometrie," *Zeitschrift für Physik* **90**, 348–354 (1934).

DiSalle, Robert. "Helmholtz's Empiricist Philosophy of Mathematics: Between Laws of Perception and Laws of Nature," in Cahan 1993, 498–521.

————. "Spacetime Theory as Physical Geometry," *Erkenntnis* **42**, 317–337 (1995).

Dombrowski, Peter. *150 Years After Gauss' Disquisitiones generales circa superfices curva.* Paris: Société Mathématique de France, 1979.

Earman, J. S., C. N. Glymour, and J. J. Stachel (editors). *Foundations of Space-Time Theories.* Minneapolis: University of Minnesota Press, 1977.

Edwards, Harold M. *Riemann's Zeta Function.* Dover, 2001.

Ehlers, J., F. A. E. Pirani, and A. Schild, "The Geometry of Free Fall and Light Propagation," in O'Raifeartaigh 1972, 63–84.

Einstein, Albert. *Geometrie und Erfahrung.* Berlin: Julius Springer, 1921.

————. "Nichteuklidische Geometrie und Physik," *Die Neue Rundschau* **36** (1), 16–20 (1925).

————. "Geometría no euclidea y física," *Revista matemática hispano-americana*, ser. 2, **1**, 72–76 (1926).

————. "Raum, Äether und Feld in der Physik," *Forum Philosophicum* **1**, 173–184 (1930).

————. *Mein Weltbild*, edited by Carl Seelig. Zürich: Europa Verlag, 1953.

————. *Relativity: The Special and General Theory.* New York: Crown Publishers, 1961.

————. *Ideas and Opinions*, translated by Sonja Bargmann. New York: Crown, 1982.

————. *ECP: The Collected Papers of Albert Einstein.* Princeton: Princeton University Press, 1987–. Note that the page references for the translations in the companion volumes for each volume of main text will be included in square brackets.

Einstein, Albert, B. Podolsky, and N. Rosen. "Can Quantum-Mechanical Description of Physical Reality Be Considered Complete?" *Physical Review* **47**, 777–780 (1935).

Euler, Leonhard. *Letters of Euler on Different Subjects in Natural Philosophy, Addressed to a German Princess*, edited by David Brewster. New York: Harper, 1837.

Everitt, C. W. F. *James Clerk Maxwell: Physicist and Natural Philosopher.* New York: Scribner's Sons, 1975.

Ewald, William (editor). *From Kant to Hilbert: A Source Book in the Foundations of Mathematics.* Oxford: Clarendon Press, 1996.

Farwell, Ruth and Christopher Knee. "The Missing Link: Riemann's 'Commentatio,' Differential Geometry and Tensor Analysis," *Historia Mathematica* **17**, 223–255 (1990).

————. "The Geometric Challenge of Riemann and Clifford," in Boi, Flament, and Salaskis 1992, 98–106.

Fauvel, John and Jeremy Gray (editors). *the History of Mathematics: A Reader.* London: Macmillan, 1987.

Friedman, Michael. "Grünbaum on the Conventionality of Geometry," in Suppes 1973, 217–233.

————. *Foundations of Space-Time Theories: Relativistic Physics and Philosophy of Science.* Princeton: Princeton University Press, 2001.

————. "Geometry as a Branch of Physics: Background and Contexts for Einstein's 'Geometry and Experience'," in Malament 2002.

Fullinwider, S. P. "Hermann von Helmholtz: The Problem of Kantian Influence," *Studies in the History and Philosophy of Science* **21**, 41–55 (1990).

Galison, Peter. "Minkowski's Space-Time: From Visual Thinking to the Absolute World," *Historical Studies in the Physical Sciences* **10**, 85–121 (1979).

————. *Einstein's Clocks, Poincaré's Maps: Empires of Time.* New York: W. W. Norton, 2003.

Gauss, Carl Friedrich. *Inaugural Lecture on Astronomy and Papers on the Foundations of Mathematics*, translated and edited by G. Waldo Dunnington. Baton Rouge: Lousiana State University Press, 1937.

————. *Werke.* Hildesheim: Georg Olms, 1973.

————. *Theory of the Motion of the Heavenly Bodies Moving about the Sun in Conic Sections: A Translation of Theoria Motus*, translated by Charles Henry Davis. Dover, 2004.

————. *General Investigations of Curved Surfaces*, translated by James Caddall Morehead and Adam Miller Hiltebeitel, edited with an introduction and notes by Peter Pesic. Dover, 2005.

Germain, Sophie. "Mémoire sur la courbure des surfaces," *Journal für Mathematik* **7**, 1–29 (1831).

Goe, George and B. L. van der Waerden, "Comments on Millers 'The Myth of Gauss Experiment on the Euclidean Nature of Physical Space," with a reply by Arthur I. Miller, in *Isis* 65, 83–87 (1974).

Golos, Ellery B. *Foundations of Euclidean and Non-Euclidean Geometry.* New York: Holt, Rinehart, and Winston, 1968.

Grattan-Guinness, Ivor (editor). *Landmark Writings in Western Mathematics, 1640–1940.* Amsterdam: Elsevier, 2005.

Gray, Jeremy. "Non-Euclidean Geometry—A Re-Interpretation," *Historia Mathematica* 6, 236–258 (1979).

————. *Linear Differential Equations and Group Theory from Riemann to Poincaré.* Boston: Birkhuser, 1986.

————. *Ideas of Space: Euclidean, Non-Euclidean, and Relativistic*, second edition. Oxford: Clarendon Press, 1989.

————. "Poincaré and Klein—Groups and Geometries," in Boi, Flament, and Salaskis 1992, 35–44.

————. "Complex Curves: Origins and Intrinsic Geometry," in Chikara 1994, 39–50.

————, (editor). *The Symbolic Universe: Geometry and Physics 1890–1930.* Oxford: Oxford University Press, 1999a.

————. "Geometry—Formalisms and Intuitions," in Gray 1999a, 58–83.

————. *János Bolyai, Non-Euclidean Geometry, and the Nature of Space.* Cambridge, Mass.: Burndy Library Publications, 2004.

————. "Bernhard Riemann, Posthumous Thesis 'On the Hypotheses Which Lie at the Foundation of Geometry' (1867)," in Grattan-Guiness 2005a, 506–520.

————. "Felix Klein's Erlangen Program, 'Comparative Considerations of Recent Geometrical Researches' (1872)," in Grattan-Guiness 2005b, 544–552.

————. "Gauss and Non-Euclidean Geometry," in Prékopa and Molnár 2006, 61–80.

Greenberg, Marvin Jay. *Euclidean and Non-Euclidean Geometries: Development and History*, second edition. New York: W. H. Freeman, 1974.

Grene, Marjorie and Debra Nails (editors). *Spinoza and the Sciences.* Dordrecht: D. Reidel, 1986.

Grünbaum, Adolf. "On the Ontology of the Curvature of Empty Space in the Geometrodynamics of Clifford and Wheeler," in Suppes 1973, 268–295.

Harrison, Edward. *Darkness at Night: A Riddle of the Universe.* Cambridge, Mass.: Harvard University Press, 1987.

Hatfield, G. *The Natural and the Normative: Theories of Space Perception from Kant to Helmholtz.* Cambridge, Mass.: MIT Press, 1990.

Hawkins, Thomas. "The Erlanger Programm of Felix Klein: Reflections on its Place in the History of Mathematics," *Historia Mathematica* **11**, 442–470 (1984).

————. *Emergence of the Theory of Lie Groups: An Essay in the History of Mathematics 1869–1926.* New York: Springer, 2000.

Heimann, P. M. "Ether and Imponderables," in Cantor and Hodge 1981, 61–83.

Heinzmann, Gerhard. "Helmholtz and Poincaré's Considerations on the Genesis of Geometry," in Boi, Flament, and Salaskis 1992, 245–249.

Helmholtz, Hermann von. "Ueber die Thoerie der zusammengesetzeten Farben" (1852), in Helmholtz 1883, 2:3–23.

————. "Ueber die Thatsachen, die der Geometrie zum Grunde liegen" (1868), in Helmholtz 1883, 2:618–639, and Helmholtz 1968, 32–60; translated in Helmholtz 1977, 39–71.

————. "The Origin and Meaning of Geometrical Axioms," *Mind* **3**, 302–321 (1876).

————. "The Origin and Meaning of Geometrical Axioms (II)," *Mind*, **3**, 212–225 (1878), also in Helmholtz 1971, 360–365.

———. *Wissenschaftliche Abhandlungen.* Leipzig: Johann Ambrosius Barth, 1883.

———. *On the Sensations of Tone as a Physiological Basis for the Theory of Music*, second English edition by Alexander J. Ellis, with an introduction by Henry Margenau. Dover, 1954.

———. *Helmholtz's Treatise on Physiological Optics*, edited by James P. C. Southall. Dover, 1962a.

———. *Popular Scientific Lectures*, introduced by Morris Kline. Dover: 1962b.

———. *Über Geometrie.* Darmstadt: Wissenschaftliche Buchgesellschaft, 1968.

———. *Selected Writings of Hermann von Helmholtz*, edited by Russell Kahl. Middletown, CT: Wesleyan University Press, 1971.

———. *Epistemological Writings*, translated by Malcolm F. Lowe, edited by Robert S. Cohen and Yehuda Elkana. Dordrecht: D. Reidel, 1977.

Henrici, O. "The Axioms of Geometry," *Nature* **29** 453–454 (1884).

Herbart, Johann Friedrich. *Sämtliche Werke*, edited by Gustav Hartenstein. Leipzig: Leopold Voss, 1850–1851.

———. *A Text-Book in Psychology: An Attempt to Found the Science of Psychology on Experience, Metaphysics, and Mathematics*, translated by Margaret K. Smith. New York: Appleton, 1897.

Hilbert, David. *Foundations of Geometry*, translated by Leo Unger, tenth edition. La Salle, Ill.: Open Court, 1959.

Holton, Gerald. *Einstein, History, and Other Passions.* Reading, Mass.: Addison-Wesley, 1996.

Hoppe, Edmund. "C. F. Gauss und der euklidische Raum." *Die Naturwissenschaften* **13**, 743–744 (1925).

Huggett, Nick (editor). *Space from Zeno to Einstein: Classic Readings with a Contemporary Commentary.* Cambridge, Mass.: MIT Press, 1999.

Ionescu-Pallas, Nicholas and Liviu Sofonea, "Bernhard Riemann: A Forerunner of Classical Electrodynamics (An Historical Epistemological Approach)," *Organon* **22/23**, 259–272 (1986/1987).

Jammer, Max. *The Conceptual Development of Quantum Mechanics.* New York, McGraw- Hill, 1966.

———. *Concepts of Space: The History and Theories of Space in Physics*, second edition. Cambridge, Mass: Harvard University Press, 1969.

Jevons, William Stanley. "Helmholtz on the Axioms of Geometry," *Nature* **4**, 481–482 (1871).

Kaiser, Walter. *Theorien der Electrodynamik im 19. Jahrhundert.* Hildesheim: Gerstenberg, 1981.

Kant, Immanuel. *Kant's Inaugural Dissertation and Early Writings on Space*, translated by John Handyside. Chicago: Open Court, 1928.

———. *Critique of Pure Reason*, translated and edited by Paul Guyer and Allen W. Wood. Cambridge: Cambridge University Press, 1998.

Klaf, A. Albert. *Trigonometry Refresher*. Dover, 2005.

Klein, Felix. "A Comparative Review of Recent Researches in Geometry" [Erlangen Program], translated by M. W. Haskell, *Bulletin of the New York Mathematical Society* **2**, 215–249 (1893); original text in Klein 1921–1923, 1:460–497.

————. *The Evanston Colloquium Lectures on Mathematics*. New York: American Mathematical Society, 1911.

————. *Gesammelte mathematische Abhandlungen*. Berlin: Springer, 1921–1923.

————. *Vorlesungen über nicht-Euklidische Geometrie*, edited by W. Rosemann. Berlin: Julius Springer, 1928.

————. *On Riemann's Theory of Algebraic Functions and their Integrals*. Dover, 1963.

————. *Development of Mathematics in the Nineteenth Century*, translated by M. Ackerman. Brookline, Mass: Math Sci Press, 1979.

Kline, Morris. *Mathematical Thought from Ancient to Modern Times*. New York: Oxford University Press, 1972.

Kramer, Edna E. *The Nature and Growth of Modern Mathematics*. Princeton: Princeton University Press, 1982.

Krüger, Lorentz (editor). *Universalgenie Helmholtz: Rückblick nach 100 Jahren*. Berlin: Akademie Verlag, 1994.

Lanczos, Cornelius. *Albert Einstein and the Cosmic World Order*. New York: Wiley, 1965.

Land, J. P. N. "Kant's Space and Modern Mathematics," *Mind* **2**, 38–46 (1877)

Laugwitz, Detlef. *Differentialgeometrie*. Stuttgart: Teubner, 1977.

————. *Bernhard Riemann 1826–1866: Turning Points in the Conception of Mathematics*, translated by Abe Shenitzer. Boston: Birkhäuser, 1999.

Leibniz, G. W. *Philosophical Essays*, edited and translated by Roger Ariew and Daniel Garber. Indianapolis: Hackett, 1989.

Lenoir, Timothy. "The Eye as Mathematician: Clinical Practice, Instrumentation, and Helmholtz's Construction of an Empiricist Theory of Vision," in Cahan 1993, 109–153.

Lie, Sophus. *Gesammelte Abhanglungen*. Leipzig: B. G. Teubner, 1935.

————. *Über die Grundlagen der Geometrie*. Darmstadt: Wissenschaftliche Buchgessellschaft, 1967.

Lietzmann, W. *Visual Topology*, translated by M. Bruckheimer. New York: Elsevier, 1965.

Lorentz, H. A., A. Einstein, H. Minkowski, and H. Weyl. *The Principle of Relativity: A Collection of Original Memoirs on the Special and General Theory of Relativity*. Dover, 1923 [the date of the original edition; the Dover edition shows no separate date].

Lützen, Jesper. "The Geometrization of Analytical Mechanics: A Pioneering Contribution by Joseph Liouville (ca. 1850)," in Rowe and McCleary 1989, 77–97.

———. "Interactions between Mechanics and Differential Geometry in the 19th Century," *Archive for the History of Exact Science* 49, 1–72 (1995).

———. "Geometrizing Configurations: Heinrich Hertz and his Mathematical Precursors," in Gray 1999a, 25–46.

MacAdam, David L. (editor). *Sources of Color Science* Cambridge, Mass.: MIT Press, 1970.

Mach, Ernst. *Space and Geometry in the Light of Physiological, Psychological and Physical Inquiry*, translated by Thomas J. McCormack. Chicago: Open Court, 1906.

Malament, David (editor). *Reading Natural Philosophy.* Chicago: Open Court, 2002.

Manning, Henry P. (editor). *The Fourth Dimension Simply Explained.* Dover, 1960.

Maurin, Krzysztof. *The Riemann Legacy: Riemannian Ideas in Mathematics and Physics.* Dordrecht: Kluwer Academic, 1997.

Maxfield, John E. and Margaret W. Maxfield, *Abstract Algebra and Solution by Radicals.* Dover, 1992.

Maxwell, James Clerk. *A Treatise on Electricity and Magnetism*, third edition. Dover, 1954.

Mazur, Barry and Peter Pesic. "On Mathematics, Imagination & the Beauty of Numbers," *Dædalus* **134**, 124–130 (2005).

Meschkowski, H. *Noneuclidean Geometry*, translated by Abe Schnitzer. New York: Academic Press, 2000.

Mill, John Stuart. *A System of Logic, Ratiocinative and Inductive; Being a Connected View of the Principles of Evidence, and the Methods of Scientific Investigation*, edited by J. M. Robson, introduction by R. F. McRae. Toronto: University of Toronto Press, 1974.

Miller, Arthur. "The Myth of Gauss' Experiment on the Euclidean Nature of Physical Space," *Isis* **63**, 345–348 (1972).

———. "Einstein, Poincaré, and the Testability of Geometry," in Gray 1999a, 47–57.

Monastyrsky, Michael. *Riemann, Topology, and Physics*, translated by James King and Victoria King. Boston: Birkhuser, 1987.

Neuenschwander, Erwin. "A Brief Report on a Number of Recently Discovered Sets of Notes on Riemann's Lectures and on the Transmission of the Riemann *Nachlass*" *Historia Mathematica* **15**, 101–113 (1988), also included in Riemann 1990, 855–867.

Newcomb, Simon. "Elementary Theorems Relating to the Geometry of a Space of Three Dimensions and of Uniform Positive Curvature in the Fourth Dimension," *Journal für die reine und angewandte Mathematik*" **83**, 293–299 (1877).

———. "Modern Mathematical Thought," *Nature* **49**, 325–329 (1894).

———. "The Philosophy of Hyper-Space," *Science* **7**, 1–7 (1898).

Newman, James R. (editor) *The World of Mathematics.* New York: Simon and Schuster, 1956.

Newton, Isaac. *The* Principia: *Mathematical Principles of Natural Philosophy,* translated by I. Bernard Cohen and Anne Whitman. Berkeley: University of California Press, 1999.

Norton, John D. "Geometries in Collision: Einstein, Klein, and Riemann," in Gray 1999a, 128–144.

Nowak, Gregory. "Riemann's *Habilitationsvortrag* and the Synthetic *A Priori* Status of Geometry," in Rowe and McCleary 1989, 17–46.

O'Raifeartaigh, L. (editor). *General Relativity: Papers in Honour of J. L. Synge.* Oxford: Clarendon Press, 1972.

Pais, Abraham. *'Subtle is the Lord ...': The Science and the Life of Albert Einstein.* Oxford: Oxford University Press, 1982.

Paty, Michel. "Physical Geometry and Special Relativity: Einstein and Poincaré," in Boi, Flament, and Salaskis 1992, 127–149.

———. "Einstein and Spinoza," in Grene and Nails 1986, w67–302.

Pesic, Peter. "The Fields of Light," *St. John's Review* **38**, 1–16 (1988–1989).

———. "The Smallest Clock," *European Journal of Physics* **14**, 90–92 (1993).

———. "Einstein and Spinoza: Determinism and Identicality Reconsidered," *Studia Spinozana* **12**, 195–203 (1996).

———. *Seeing Double: Shared Identities in Physics, Philosophy, and Literature.* Cambridge, Mass.: MIT Press, 2002.

———. *Abel's Proof: An Essay on the Sources and Meaning of Mathematical Unsolvability.* Cambridge, Mass.: MIT Press, 2003.

Pesic, Peter and Stephen P. Boughn. "The Weyl-Cartan theorem and the naturalness of general relativity," *European Journal of Physics* **24**, 261–266 (2003).

Pettoello, Renato. "Dietro la superfice dei fenomeni. Frammenti di filosofia in Bernhard Riemann," *Rivista di storia della filosofia* **4**, 697–728 (1988).

Piaget, Jean. *Psychology and Epistemology: Towards a Theory of Knowledge,* translated by Arnold Rosin. New York: Viking, 1971.

Poincaré, Henri. "Sur les hypothèses fondamentales de la géométrie," in Poincaré 1956, 11:79–91.

———. "Les géométries non Euclidiennes," *Revue générale des Sciences pures et appliquées,* **2**, 669–774 (1891).

———. "Correspondance sur les géométries non Euclidiennes," *Revue générale des Sciences pures et appliquées,* **3**, 74–75 (1892a).

———. "Non-Euclidean Geometry," *Nature* **45**, 404–407 (1892b) [translation of Poincaré 1891].

———. "On the foundations of geometry," *Monist* **9**, 1–43 (1898); also in Ewald 1996, 2:982–1011.

———. *Science and Hypothesis.* Dover, 1952.

———. *Œuvres de Henri Poincaré.* Paris: Gauthier-Villars, 1956.

Poggi, Stefano and Maurizio Bossi (editors). Romanticism in Science: Science in Europe, 1790-1840. Dordrecht : Kluwer Academic, 1994.

Portnoy, Esther. "Riemann's Contribution to Differential Geometry," *Historia Mathematica* **8**, 1–18 (1982).

Prékopa, András. "The Revolution of János Bolyai," in Prékopa and Molnár 2006, 3–59.

Prékopa, András and Emil Molnár (editors). *Non-Euclidean Geometries: János Bolyai Memorial Volume.* New York: Springer, 2006.

Rédei, L. *Foundations of Euclidean and Non-Euclidean Geometries According to F. Klein.* Oxford: Pergamon, 1968.

Reich, Karin. Die Geschichte der Differentialgeometrie von Gauss bis Riemann (1828-1868). Archive for History of Exact Science 11, 273-382 (1973).

Reichenbach, Hans. *The Philosophy of Space and Time*, translated by Maria Reichenbach and John Freund with introductory remarks by Rudolf Carnap. Dover, 1958.

Richards, Joan L. "The Evolution of Empiricism: Hermann von Helmholtz and the Foundations of Geometry," *British Journal for the Philosophy of Science* **28**, 235–253 (1977).

————. *Mathematical Visions: The Pursuit of Geometry in Victorian English.* Boston: Academic Press, 1988.

Riemann, Bernhard. "On the Hypotheses Which Lie at the Bases of Geometry," translated by W. K. Clifford. *Nature* **8**, 114–117, 136–137 (1873).

————. *Ueber die Hypotheses, welche der Geometrie zu Grunde liegen*, edited by H. Weyl. Berlin: Springer, 1919.

————. *Gesammelte Mathematische Werke und Wissenschaftlicher Nachlass*, edited by Heinrich Weber and Richard Dedekind, with an introduction by Hans Lewy. Dover: 1953.

————. *Gesammelte Mathematische Werke, Wissenschaftlicher Nachlass, und Nachträge*, edited by Raghavan Narasimhan after the edition by Heinrich Weber and Richard Dedekind. Berlin: Springer-Verlag, 1990.

————. *Sulle ipotesi che stanno alla base della geometria e altri scritti scientifici e filosofici*, edited by Renato Pettoello. Torino: Bollati Boringhieri, 1994.

Rindler, Wolfgang. "General Relativity Before Special Relativity: An Unconventional Overview of Relativity Theory," *American Journal of Physics* **62**, 887–893 (1994).

Rosenfeld, B. A. *A History of Non-Euclidean Geometry: Evolution of the Concept of a Geometric Space*, translated by Abe Shenitzer. New York: Springer-Verlag, 1988.

Rowe, David E. "Felix Klein's 'Erlanger Antrittsrede': A Transcription with English Translation and Commentary," *Historia Mathematics* **12**, 123–141 (1985)

————. "Klein, Lie, and the 'Erlanger Programm'," in Boi, Flament, and Salaskis 1992, 45–54.

Rowe, David E. and John McCleary (editors). *The History of Modern Mathematics.* New York: Academic Press, 1989.

Russell, Bertrand A. W. *An Essay on the Foundations of Geometry*, with a foreword by Morris Kline. Dover, 1956.

Salecker, H. and E. P. Wigner, "Quantum Limitations of the Measurement of Space-Time Distances," *Physical Review* **109**, 571–577 (1958).

Schlick, Moritz. *Space and Time in Contemporary Physics: An Introduction to the Theory of Relativity and Gravitation*, translated by H. L. Brose with an introduction by F. A. Lindemann. Dover, 1963.

Scholtz, Erhard. *Geschichte des Mannigfaltigkeitsbegriff von Riemann bis Poincaré.* Basel: Birkhäuser. 1980.

————. "Herbart's Influence on Bernhard Riemann," *Historia Mathematica* **9**, 413–440 (1982a).

————. "Riemanns frühe Notizen zum Mannigfaltigkeitsbegriff und zu den Grundlagen der Geometrie," *Archive for History of Exact Sciences* **27**, 213–232 (1982b).

————. "Riemann's Vision of a New Approach to Geometry," in Boi, Flament, and Salaskis 1992, 22–34.

————. "Weyl and the Theory of Connections," in Gray 1999a, 260–284.

Schüller, Volkmar. "Das Helmholtz-Liesche Raumproblem und seine ersten Lösungen," in Krüger 1994, 260–275.

Sklar, Lawrence. *Space, Time, and Spacetime.* Berkeley: University of California Press, 1977.

Smith, David Eugene (editor). *A Source Book in Mathematics.* Dover, 1959.

Solovine, Maurice (editor). *Albert Einstein: Lettrès à Maurice Solovine.* Paris: Gauthiers-Villars, 1956.

Speiser, Andreas. "Naturphilosophische Untersuchungen von Euler und Riemann," *Journal für die reine und angewandte Mathematik* **157**, 105–114 (1927).

Spivak, Michael. *A Comprehensive Introduction to Differential Geometry*, third edition. Houston: Publish or Perish, 1999.

Stein, Howard. "Some Philosophical Prehistory of General Relativity," in Earman et al. 1977.

Stillwell, John (editor). *Sources of Hyperbolic Geometry.* Providence, Rhode Island: American Mathematical Society, 1996.

Struik, Dirk J. "Schouten, Levi-Civita, and the Emergence of Tensor Calculus," in Rowe and McCleary 1989, 99–105.

Suppes, Patrick (editor). *Space, Time, and Geometry.* Dordrecht: D. Reidel, 1973.

Tazzioli, Rossana. "Fisica e 'filosofia naturale' in Riemann," *Nuncius* **8**, 105–120 (1993).

————. *Beltrami e i matematici relativistici: la meccanica in spazi curvi nella seconda metà dell'Ottocento.* Bologna: Pitagora, 2000.

————. "Riemann: Le géometre de la nature." *Pour la science* 12 (2002).

————. "Towards a History of Geometric Foundations of Mathematics: Late 19th Century," *Revue de Synthèse* **124**, 11–41 (2003).

Torretti, Roberto. *Philosophy of Geometry from Riemann to Poncaré.* Dordrecht: D. Reidel, 1978.

Turner, R. Steven. "Consensus and Controversy: Helmholtz on the Visual Perception of Space," in Cahan 1993, 154–204.

Van Fraasen, B. C. *An Introduction to the Philosophy of Time and Space.* New York: Random House, 1970.

Vuillemin, Jules. "Poincaré's Philosophy of Space," in Suppes 1973, 159–177.

Wall, C. T. C. A Geometric Introduction to Topology. Dover, 1993.

Walter, Scott. "The Non-Euclidean Style of Minkowskian Relativity," in Gray 1999a, 91–127.

Weil, André. "Riemann, Betti, and the Birth of Topology," *Archive for History of Exact Science* **20**, 91–96 (1979a).

————. "A Postscript to my Article 'Riemann, Betti, and the Birth of Topology'," *Archive for History of Exact Science* **20**, 387 (1979b).

Weyl, Hermann. *Philosophy of Mathematics and Natural Science*, translated by Olaf Helmer. Princeton: Princeton University Press, 1949.

————. *Space-Time-Matter*, translated by Henry L. Brose. Dover, 1952.

————. *The Concept of a Riemann Surface*, translated by Gerald R. Maclane. Reading, Mass.: Addison-Wesley, 1955.

————. *Mathematische Analyse des Raymproblems/Was ist Materie?* Darmstadt: Wissenschaftliche Buchgesellschaft, 1977.

————. *Symmetry.* Princeton: Princeton University Press, 1980.

————. *Riemanns geometrische Ideen, ihre Auswirkung und ihre Verknüpfung mit der Gruppentheorie (1925)*, edited by K. Chandrasekharan. Berlin: Springer-Verlag, 1988.

Wheeler, John Archibald. *Einsteins Vision: wie steht es heute mit Einsteins Vision alles als Geometrie aufzufassen?*. Berlin: Springer-Verlag, 1968.

————. *Geometrodynamics.* New York: Academic Press, 1962.

Williams, L. Pearce. *The Origins of Field Theory.* Lanham, MD: University Press of America, 1980.

Wise, Norton M. "German Concepts of Force, Energy, and the Electromagnetic Ether: 1845–1880," in Cantor and Hodge 1981.

Witten, Edward. "Reflections on the Fate of Spacetime," *Physics Today* **49**(4), 24–30 (1996), also included in Callender and Huggett 2001, 125–137.

Wussing, Hans. *The Genesis of the Abstract Group Concept*, translated by Abe Schenitzer. Cambridge, Mass.: MIT Press, 1984.

Yaglom, I. M. *Felix Klein and Sophus Lie: Evolution of the Idea of Symmetry in the Nineteenth Century.* Boston: Birkhäuser, 1988.

Zimmerman, E. J. "The Macroscopic Nature of Space-Time," *American Journal of Physics* **30**, 97–105 (1962).

Zund, J. D. "Some Comments on Riemann's Contributions to Differential Geometry," *Historia Mathematica* **10**, 84–89 (1983).

Index

A CATALOG OF SELECTED

DOVER BOOKS

IN SCIENCE AND MATHEMATICS

Mathematics

FUNCTIONAL ANALYSIS (Second Corrected Edition), George Bachman and Lawrence Narici. Excellent treatment of subject geared toward students with background in linear algebra, advanced calculus, physics and engineering. Text covers introduction to inner-product spaces, normed, metric spaces, and topological spaces; complete orthonormal sets, the Hahn-Banach Theorem and its consequences, and many other related subjects. 1966 ed. 544pp. 6⅛ x 9¼. 0-486-40251-7

ASYMPTOTIC EXPANSIONS OF INTEGRALS, Norman Bleistein & Richard A. Handelsman. Best introduction to important field with applications in a variety of scientific disciplines. New preface. Problems. Diagrams. Tables. Bibliography. Index. 448pp. 5⅜ x 8½. 0-486-65082-0

VECTOR AND TENSOR ANALYSIS WITH APPLICATIONS, A. I. Borisenko and I. E. Tarapov. Concise introduction. Worked-out problems, solutions, exercises. 257pp. 5⅝ x 8¼. 0-486-63833-2

AN INTRODUCTION TO ORDINARY DIFFERENTIAL EQUATIONS, Earl A. Coddington. A thorough and systematic first course in elementary differential equations for undergraduates in mathematics and science, with many exercises and problems (with answers). Index. 304pp. 5⅜ x 8½. 0-486-65942-9

FOURIER SERIES AND ORTHOGONAL FUNCTIONS, Harry F. Davis. An incisive text combining theory and practical example to introduce Fourier series, orthogonal functions and applications of the Fourier method to boundary-value problems. 570 exercises. Answers and notes. 416pp. 5⅜ x 8½. 0-486-65973-9

COMPUTABILITY AND UNSOLVABILITY, Martin Davis. Classic graduate-level introduction to theory of computability, usually referred to as theory of recurrent functions. New preface and appendix. 288pp. 5⅜ x 8½. 0-486-61471-9

ASYMPTOTIC METHODS IN ANALYSIS, N. G. de Bruijn. An inexpensive, comprehensive guide to asymptotic methods–the pioneering work that teaches by explaining worked examples in detail. Index. 224pp. 5⅜ x 8½ 0-486-64221-6

APPLIED COMPLEX VARIABLES, John W. Dettman. Step-by-step coverage of fundamentals of analytic function theory–plus lucid exposition of five important applications: Potential Theory; Ordinary Differential Equations; Fourier Transforms; Laplace Transforms; Asymptotic Expansions. 66 figures. Exercises at chapter ends. 512pp. 5⅜ x 8½. 0-486-64670-X

INTRODUCTION TO LINEAR ALGEBRA AND DIFFERENTIAL EQUATIONS, John W. Dettman. Excellent text covers complex numbers, determinants, orthonormal bases, Laplace transforms, much more. Exercises with solutions. Undergraduate level. 416pp. 5⅜ x 8½. 0-486-65191-6

RIEMANN'S ZETA FUNCTION, H. M. Edwards. Superb, high-level study of landmark 1859 publication entitled "On the Number of Primes Less Than a Given Magnitude" traces developments in mathematical theory that it inspired. xiv+315pp. 5⅜ x 8½. 0-486-41740-9

CALCULUS OF VARIATIONS WITH APPLICATIONS, George M. Ewing. Applications-oriented introduction to variational theory develops insight and promotes understanding of specialized books, research papers. Suitable for advanced undergraduate/graduate students as primary, supplementary text. 352pp. 5⅜ x 8½. 0-486-64856-7

COMPLEX VARIABLES, Francis J. Flanigan. Unusual approach, delaying complex algebra till harmonic functions have been analyzed from real variable viewpoint. Includes problems with answers. 364pp. 5⅜ x 8½. 0-486-61388-7

AN INTRODUCTION TO THE CALCULUS OF VARIATIONS, Charles Fox. Graduate-level text covers variations of an integral, isoperimetrical problems, least action, special relativity, approximations, more. References. 279pp. 5⅜ x 8½. 0-486-65499-0

COUNTEREXAMPLES IN ANALYSIS, Bernard R. Gelbaum and John M. H. Olmsted. These counterexamples deal mostly with the part of analysis known as "real variables." The first half covers the real number system, and the second half encompasses higher dimensions. 1962 edition. xxiv+198pp. 5⅜ x 8½. 0-486-42875-3

CATASTROPHE THEORY FOR SCIENTISTS AND ENGINEERS, Robert Gilmore. Advanced-level treatment describes mathematics of theory grounded in the work of Poincaré, R. Thom, other mathematicians. Also important applications to problems in mathematics, physics, chemistry and engineering. 1981 edition. References. 28 tables. 397 black-and-white illustrations. xvii + 666pp. 6⅛ x 9¼. 0-486-67539-4

INTRODUCTION TO DIFFERENCE EQUATIONS, Samuel Goldberg. Exceptionally clear exposition of important discipline with applications to sociology, psychology, economics. Many illustrative examples; over 250 problems. 260pp. 5⅜ x 8½. 0-486-65084-7

NUMERICAL METHODS FOR SCIENTISTS AND ENGINEERS, Richard Hamming. Classic text stresses frequency approach in coverage of algorithms, polynomial approximation, Fourier approximation, exponential approximation, other topics. Revised and enlarged 2nd edition. 721pp. 5⅜ x 8½. 0-486-65241-6

INTRODUCTION TO NUMERICAL ANALYSIS (2nd Edition), F. B. Hildebrand. Classic, fundamental treatment covers computation, approximation, interpolation, numerical differentiation and integration, other topics. 150 new problems. 669pp. 5⅜ x 8½. 0-486-65363-3

THREE PEARLS OF NUMBER THEORY, A. Y. Khinchin. Three compelling puzzles require proof of a basic law governing the world of numbers. Challenges concern van der Waerden's theorem, the Landau-Schnirelmann hypothesis and Mann's theorem, and a solution to Waring's problem. Solutions included. 64pp. 5⅜ x 8½. 0-486-40026-3

THE PHILOSOPHY OF MATHEMATICS: AN INTRODUCTORY ESSAY, Stephan Körner. Surveys the views of Plato, Aristotle, Leibniz & Kant concerning propositions and theories of applied and pure mathematics. Introduction. Two appendices. Index. 198pp. 5⅜ x 8½. 0-486-25048-2

Physics

OPTICAL RESONANCE AND TWO-LEVEL ATOMS, L. Allen and J. H. Eberly. Clear, comprehensive introduction to basic principles behind all quantum optical resonance phenomena. 53 illustrations. Preface. Index. 256pp. 5⅜ x 8½. 0-486-65533-4

QUANTUM THEORY, David Bohm. This advanced undergraduate-level text presents the quantum theory in terms of qualitative and imaginative concepts, followed by specific applications worked out in mathematical detail. Preface. Index. 655pp. 5⅜ x 8½. 0-486-65969-0

ATOMIC PHYSICS (8th EDITION), Max Born. Nobel laureate's lucid treatment of kinetic theory of gases, elementary particles, nuclear atom, wave-corpuscles, atomic structure and spectral lines, much more. Over 40 appendices, bibliography. 495pp. 5⅜ x 8½. 0-486-65984-4

A SOPHISTICATE'S PRIMER OF RELATIVITY, P. W. Bridgman. Geared toward readers already acquainted with special relativity, this book transcends the view of theory as a working tool to answer natural questions: What is a frame of reference? What is a "law of nature"? What is the role of the "observer"? Extensive treatment, written in terms accessible to those without a scientific background. 1983 ed. xlviii+172pp. 5⅜ x 8½. 0-486-42549-5

AN INTRODUCTION TO HAMILTONIAN OPTICS, H. A. Buchdahl. Detailed account of the Hamiltonian treatment of aberration theory in geometrical optics. Many classes of optical systems defined in terms of the symmetries they possess. Problems with detailed solutions. 1970 edition. xv + 360pp. 5⅜ x 8½. 0-486-67597-1

PRIMER OF QUANTUM MECHANICS, Marvin Chester. Introductory text examines the classical quantum bead on a track: its state and representations; operator eigenvalues; harmonic oscillator and bound bead in a symmetric force field; and bead in a spherical shell. Other topics include spin, matrices, and the structure of quantum mechanics; the simplest atom; indistinguishable particles; and stationary-state perturbation theory. 1992 ed. xiv+314pp. 6⅛ x 9¼. 0-486-42878-8

LECTURES ON QUANTUM MECHANICS, Paul A. M. Dirac. Four concise, brilliant lectures on mathematical methods in quantum mechanics from Nobel Prize-winning quantum pioneer build on idea of visualizing quantum theory through the use of classical mechanics. 96pp. 5⅜ x 8½. 0-486-41713-1

THIRTY YEARS THAT SHOOK PHYSICS: THE STORY OF QUANTUM THEORY, George Gamow. Lucid, accessible introduction to influential theory of energy and matter. Careful explanations of Dirac's anti-particles, Bohr's model of the atom, much more. 12 plates. Numerous drawings. 240pp. 5⅜ x 8½. 0-486-24895-X

ELECTRONIC STRUCTURE AND THE PROPERTIES OF SOLIDS: THE PHYSICS OF THE CHEMICAL BOND, Walter A. Harrison. Innovative text offers basic understanding of the electronic structure of covalent and ionic solids, simple metals, transition metals and their compounds. Problems. 1980 edition. 582pp. 6⅛ x 9¼. 0-486-66021-4

CATALOG OF DOVER BOOKS

A TREATISE ON ELECTRICITY AND MAGNETISM, James Clerk Maxwell. Important foundation work of modern physics. Brings to final form Maxwell's theory of electromagnetism and rigorously derives his general equations of field theory. 1,084pp. 5⅜ x 8½. Two-vol. set. Vol. I: 0-486-60636-8 Vol. II: 0-486-60637-6

QUANTUM MECHANICS: PRINCIPLES AND FORMALISM, Roy McWeeny. Graduate student-oriented volume develops subject as fundamental discipline, opening with review of origins of Schrödinger's equations and vector spaces. Focusing on main principles of quantum mechanics and their immediate consequences, it concludes with final generalizations covering alternative "languages" or representations. 1972 ed. 15 figures. xi+155pp. 5⅜ x 8½. 0-486-42829-X

INTRODUCTION TO QUANTUM MECHANICS With Applications to Chemistry, Linus Pauling & E. Bright Wilson, Jr. Classic undergraduate text by Nobel Prize winner applies quantum mechanics to chemical and physical problems. Numerous tables and figures enhance the text. Chapter bibliographies. Appendices. Index. 468pp. 5⅜ x 8½. 0-486-64871-0

METHODS OF THERMODYNAMICS, Howard Reiss. Outstanding text focuses on physical technique of thermodynamics, typical problem areas of understanding, and significance and use of thermodynamic potential. 1965 edition. 238pp. 5⅜ x 8½. 0-486-69445-3

THE ELECTROMAGNETIC FIELD, Albert Shadowitz. Comprehensive undergraduate text covers basics of electric and magnetic fields, builds up to electromagnetic theory. Also related topics, including relativity. Over 900 problems. 768pp. 5⅝ x 8¼. 0-486-65660-8

GREAT EXPERIMENTS IN PHYSICS: FIRSTHAND ACCOUNTS FROM GALILEO TO EINSTEIN, Morris H. Shamos (ed.). 25 crucial discoveries: Newton's laws of motion, Chadwick's study of the neutron, Hertz on electromagnetic waves, more. Original accounts clearly annotated. 370pp. 5⅜ x 8½. 0-486-25346-5

EINSTEIN'S LEGACY, Julian Schwinger. A Nobel Laureate relates fascinating story of Einstein and development of relativity theory in well-illustrated, nontechnical volume. Subjects include meaning of time, paradoxes of space travel, gravity and its effect on light, non-Euclidean geometry and curving of space-time, impact of radio astronomy and space-age discoveries, and more. 189 b/w illustrations. xiv+250pp. 8⅜ x 9¼. 0-486-41974-6

STATISTICAL PHYSICS, Gregory H. Wannier. Classic text combines thermodynamics, statistical mechanics and kinetic theory in one unified presentation of thermal physics. Problems with solutions. Bibliography. 532pp. 5⅜ x 8½. 0-486-65401-X

Paperbound unless otherwise indicated. Available at your book dealer, online at **www.doverpublications.com**, or by writing to Dept. GI, Dover Publications, Inc., 31 East 2nd Street, Mineola, NY 11501. For current price information or for free catalogues (please indicate field of interest), write to Dover Publications or log on to **www.doverpublications.com** and see every Dover book in print. Dover publishes more than 500 books each year on science, elementary and advanced mathematics, biology, music, art, literary history, social sciences, and other areas.